ORGANIC FARMING FOR SUSTAINABLE AGRICULTURE

SECOND ENLARGED EDITION

A. K. DAHAMA
Professor of Agronomy
College of Agriculture
Rajasthan Agricultural University, Bikaner

AGROBIOS

Published by:
AGROBIOS (INDIA)
Agro House, Behind Nasrani Cinema
Chopasani Road, Jodhpur 342 003
Phone: 91-0291-2643993
Website: *www.agrobiosonline.com*
E. mail: *agrobiosindia@gmail.com*

AGROBIOS

First Published: 1996
Second Edition: 19997

ISBN No.: 978-81-7754-058-1

Published by: Dr. Updesh Purohit for Agrobios (India), Jodhpur
Laser Typeset at: Yashee Computers, Jodhpur
Cover Design by: Shyam
Printed at: Agrobios Digitals, Jodhpur

Preface to Second Edition

The first edition of Organic Farming for Sustainable Agriculture, released in 1996, has been well received. Organic fanning systems have drawn attention of a large section of planners and scientists at the global level. A number of conferences are being organised the world over to take a stock of the impact of the chemical agriculture on food production and natural resources; and also to give a systematic approach to organic fanning systems for safeguarding the interests of both the present and the future generations.

In the present edition, a chapter on Biogas Technology for Organic Farming and recently generated information on various aspects of organic farming with special reference to ecological agriculture, pollution from agrochemicals, etc. have been added.

July, 1997 **A. K. DAHAMA**

Preface to First Edition

With the increase in crop yields from modem farming techniques reaching a plateau in most of the countries, and the environmental problems due to excessive use of chemical fertilizers and pesticides becoming a matter of concern, the need for sustainable agriculture is increasingly being felt, the world over. In the process of attaining higher levels of food production for matching, the demands of the growing population during the past four decades, emphasis was laid on intensive agricultural practices. The technologies generated over the past four decades have proved. detrimental to the natural resource base and the environment in different parts of the world.

The gravity of the environmental degradation has drawn the attention of the scientists and planners towards finding out ecologically sound, viable and sustainable farming systems for different agroclimatic situations, keeping in view the needs of the present and future generations. Steps taken in this direction demand a global policy, as environmental degradation has far reaching and long-term effects. The farming practices that are governed by the principles of ecology have proved to be highly productive and sustainable in several parts of the world. Adoption of such farming systems should not be treated as a reversion to the inefficient and less productive agricultural practices of the era prior to that of intensive agriculture.

In view of resurgence of interest in alternate agriculture in recent years, organic farming has been considered to be a sound and viable option in most of the countries. In this direction the recommendations of the Atlanta Conference of 1981 on "Organic Farming' have acted as catalysts in triggering interest in the organic agricultural systems across the world. The available information on different aspects of organic farming is very meagre and fragmented. Some of the research work undertaken during the era prior to the adoption of the intensive agricultural systems, has relevance to the present day Organic Farming. Research information generated during the past four decades of high-tech agriculture has little appreciation for organic farming systems, specially those related to sustainable agriculture. Thus, the progress of the organic farming systems will largely depend on generation of new technology, suitable to a particular structural and agroclimatic set-up.

Here, an attempt has been made to compile available information and present the current status of organic farming, its strength and weaknesses, and research needs, for providing a basis for development of alternate farming systems, which are sustainably sound and viable on a long-term basis. The text covers available information on principles and practices of organic farming, status of sustainable agricultural systems, agricultural pollution through soil and agrochemicals, traditional and non- traditional -additives, on-farm and off-farm wastes and their use, and future trends in respect of organic farming.

Agricultural scientists endorse all the beneficial aspects of organic farming practices but, at the same time, express doubts about their adoption on a large scale, in view of lack of adequate research support. Thus, adoption of organic farming practices should be gradual and systematic, backed-up by a sound research and development net-work. The present world scenario demands that all nations become invested in the greatest challenge ever faced by any sentient species-sustaining natural resources and agriculture for supporting life on the planet. Organic agriculture has an important role to play in this direction. I hope this publication shall be found useful in meeting some of the requirements of the noble cause as mentioned above.

I take this opportunity to express gratitude to my friend, Dr. S. S. Purohit, who was instrumental in encouraging me to present information on such an important issue. Thanks are also extended to Mr. Updesh Purohit of M/s Agro- Botanical Publishers (India), Bikaner for necessary assistance **provided in bringing out this book.** My wife, Daisy, and daughters, Nandita and Namrata, who provided required conditions at home to complete this work, deserve special appreciation.

A. K. DAHAMA

Contents

CHAPTER 1

Organic Farming: An Overview

The development of industrialized farming paralleled technological advances, and was especially encouraged by improved analytical methods and advances in various disciplines of agricultural sciences. This tandem development led to the belief that modern agricultural developments were progressively inevitable and must be beneficial from a socio-economic standpoint. The quantitative viewpoint brought to bear on agriculture was important, but in the process, quality was overlooked. Negative environmental side effects associated with industrial farming were considered to be unimportant.

Increasing the crop yields meant intensification in the use of fertilizers and pesticides in plant production and the increased use of concentrates in feeds and medicinal treatments in animal production. The consequence of intensification has been a reduction in the plant and animal species dependent upon local conditions. In animal husbandry, intensification has meant increased use of concentrates and an increase in the number of animals per hectare. To achieve this the productive capacity of crops and soils was increased. A decrease in the number of plant species in pastures was the consequence of intensive fertilization. In some areas, only one or two high-yielding grass species were grown. Besides negative environmental impacts, intensification also had an influence on soil and animal fertility- a problem which is of serious concern to the farmers the world over.

IMPACT OF HIGH TECHNOLOGY AGRICULTURE ON CROP PRODUCTION AND ENVIRONMENT

Modern agriculture largely depends on the use of fossil fuel-based inputs such as chemical fertilizers, pesticides, herbicides and labour saving but energy intensive farm machinery. While the application of such high input technologies has undoubtedly increased production and labour efficiency, there is a growing concern over their adverse effects on soil productivity and environmental quality which is beginning to recognize that the farmer has a great social responsibility as a land owner than merely agribusiness considerations.

Adverse Effects of Agricultural Practices on the Farm

Although high-input agriculture has certainly resulted in spectacular gains in productivity, it has degraded some of the natural bases on which this system rests. The ultimatc costs of this degradation on the farm are borne by the farmers themselves and by society at large, be they in lowered productivity, environmental clean-up, or further technology to solve a technologically induced problem. Some of these on-farm impacts are mentioned here.

Soil Erosion

During the late 1970s, the USDA informed that: up to half of the crop-land in the USA showed serious erosion of its precious, nutrient-rich topsoil. The numbers are hard for a lay person to comprehend; over 15 tons per hectare of topsoil lost, on average. Global figures are just as alarming: the United Nations is using satellites to track soil erosion, and estimates that in northern India, losses per hectare exceed those of the United States. The effect of these losses of topsoil on the farm is devastating. As wind and water remove the finer particles, the subsoil that is left is coarser, with less organic matter to break up the soil, and more fertilizers are needed to make it support plant growth. Ultimately, eroded soils can become useless for crop production.

Water Availability

Water is an essential component for plant growth, and in arid and semiarid regions, the availability of water for agriculture must be improved through irrigation. According to the FAO, 270 million ha in the world are irrigated, and this consumes 70% of the world's total human-induced water consumption.

About half the water used in the United States for irrigation comes from underground aquifers (for example, the Ogallala aquifer in the Southwest), from which it is pumped to the surface. The same is true in the Indo-Gangetic plain of India. These aquifers are slowly and constantly recharged through rain percolation in the soil. But the rate at which they are being depleted is much faster than slow recharge, and the level of the water table is falling. Mainly in the view of the demand of the intensive aquatic systems to water requirements.

Different crop plants have different needs for water. When this is expressed in economic terms, the cash value of crop per volume of water used is seen to vary greatly. Yet governments often have "cheap water" policies, providing subsidized water to those farmers who grow crops that are water hungry in locations that are water proof. This promotes depletion of water resources.

Salinization

Irrigation of salt-rich arid and semi-arid lands has an added environmental cost. When water use is excessive, the underground water table rises and brings with it dissolved salts from the lower regions of the soil. This initially lowers crop yield, and if the water evaporates and leaves salts caked on the surface, ultimately renders the soil agronomically useless unless drastic and expensive measures are taken to remove the salts. The extent of salinization is more difficult to measure than simple soil erosion, because the effects of salinized soil are underground, at least initially. Estimates suggest that in Bangladesh river delta salinization is significantly reducing yields in about one-third of the crop-land, while in the desert of the Southwest U.S. the figure approaches one-fourth.

Fertilizer and Pesticide Contamination

Applying nitrogen in fertilizer in excess of the amount removed each year with the crop can lead to loss of nitrogen from the soil environment. Because nitrate is negatively charged, it does not bind to negatively charged soil particles and so is poorly stored in the soil. When soil water exceeds plant needs (through rain or irrigation), nitrate leaches into soil ground water. This can contaminate the wells, and nitrates in drinking water pose a health problem to the animals and people who use such wells.

The extensive use of pesticides poses a more direct problem for farm workers. Because these people are generally outside of the main-stream of health care estimates of the extent of pesticide induced health problems are fraught with difficulty. But incidents of acute poisoning certainly occur with alarming frequency, and epidemiological data show that workers who handle pesticides more than 20 days a year have an increased risk of developing certain types of cancer.

Genetic Erosion

The demands of modern agriculture have led to a narrowing of the genetic base of crop plants. The initial success of the newly bred and selected strains has repeatedly led to their rapid adoption by all farmers, and abandonment of the richly diverse land-races and wild relatives. For example, in the United States six cultivars (out of tens of thousands worldwide) account for a third of the wheat planted. In Greece, virtually all land-races of wheat have disappeared over the past 40 years. The dangers of such a narrow genetic base are obvious in terms of susceptibility to pests (for example, the tungro virus that devastated IR-8 rice in the Philippines during the 1970s).

Adverse Effects of Agricultural Practices Away from the Farm

In addition to the serious effects of high-intensity farming in degrading the resource base for future farming, the environment away from the farm can be adversely affected in the following ways:

Soil Erosion

Eroded soil particles, especially carried by water, often end up filling reservoirs, increasing the costs of water extraction, and clogging navigable waterways. According to the United States Department of Agriculture, about 1 billion tons of eroded soils are deposited in water ways each year in the USA. The removal of 10 per cent of the highly eroded lands from cultivation currently would reduce this deposition by 20 per cent. Most difficult to cultivate without erosion are hillsides. Hillside cultivation requires careful management if soil erosion is to be avoided.

Fertilizer and Pesticide Contamination

Farming is a major source of water pollution by fertilizers and pesticides dissolved or suspended in soil water. Excessive nitrogenous fertilization can contaminate drinking water. It is estimated that 25 per cent of the people living in Europe are drinking water with nitrates above recommended levels (25 mg per litre). Pesticide contamination of groundwater, and then drinking water, is also a potentially serious problem. Although the effects of DDT on certain bird populations are well known, the effects of low concentrations of other pesticides on people have not been established. In the USA, detectable levels of 39 pesticides have been observed in ground water in 34 states. Although in most cases the levels are below those deemed unsafe, long-term effects of this exposure are not known.

Socio-Economic Values

Until recently in human history, there was enough land and associated resources to meet the needs of rural populations around the world. But now the easily converted land and easily extractable resources are already being used for agriculture, and an increasing number of poor people lack the economic resources to exploit the available land for optimal crop growth.

The cycle of resource degradation, hunger, and poverty can best be illustrated by considering a rural family living in northern India. Conversion of forest lands to farming, and the needs of a growing population for wood for cooking and heating, have led to a serious decline in the availability of fuel wood. Animal manure and crop residues may be used instead of wood for fuel; but this eliminates them from the soil as fertilizer, with concomitant reductions in crop yields and increases in erosion. If the local market is used to buy fuel wood, the price may be high and this results in less money available to purchase agricultural inputs; once again, yields fall and the lands are not cared for.

Ultimately, the small farmer must make difficult decisions, and it is a no-win situation. In the struggle to survive, long-term considerations of land productivity are far from paramount. Many subsistence farmers in developing countries face a dilemma. Either they must eke out a living with dignity on marginal land, thereby contributing to deforestation and land degradation, or they must go to the rapidly growing cities where there are few jobs and where life itself is degrading for so many.

Economic and Political Issues

Two objectives have driven the agricultural policies of the developed countries:

1. To provide adequate food for the people
2. To provide adequate incomes for farmers

Left unstated but surely considered are such aspects as the preservation of the family farm, foreign trade balances, the incomes of ancillary industries that provide inputs, political voting blocs, and so on. And left almost un-thought are such things as nutrition and nutritional diseases, the distribution of food to all the population, and the environmental costs of agricultural production.

In the USA, the policy that has the greatest impact on farming is the Farm Law, a commodity programme under which the federal government guarantees the farmer a minimum price for crops produced on the farm. Two-thirds of all U.S. crop-lands are now enrolled in commodity programmes, with 80-95 per cent of the acreage planted to rice, maize, wheat, sorghum and cotton. Total payments are about $25 billion per year.

Farming in the United States has become a taxpayer-supported industry, where the government contracts for production, buys products, and indirectly pays workers. As a result of the commodity programme, farmers cannot easily introduce crop rotation, because they have no "authorization" to grow the crops that might fit in a rotation schedule. The 1990 revision of the Farm Bill began to deal with this, and allows a supported farmer to plant up to 25 per cent of the acreage with other crops without losing acreage-based supports.

In Japan, the desire for local production of rice rests not only on tradition, but also on the historically strong tendency for self-sufficiency. With China as an unreliable trading partner and strong local political pressures to support rice farmers, the Government of Japan has the highest price supports of the industrialized world.

The subsidy programme also encourages the excessive use of fertilizers and pesticides, to attain high crop yields. If yields are high, the farmer safeguards future financial returns, even if it costs more to produce the crop than it is worth. According to the law of diminishing returns, it takes much more fertilizer and other inputs to raise production from 6 to 8 tons per hectare, than from 4 to 6 tons per hectare. Pushing production up to reach such a high level may not be cost-effective but the tax-payers foot the bill. However, the large amounts of fertilizer and pesticide required to reach this production level put an unnecessarily heavy load on the environment.

The General Agreement on Tariffs and Trade (GATT) governs trade between many countries of this world. In the recent negotiations of the Uruguay Round it is not surprising that a major stumbling block was agriculture. The United States put forward a bold proposal to eliminate tariffs and subsidies on agricultural products altogether. But this was rejected by the European countries, for domestic political reasons. Nevertheless, these negotiations were the beginning of the process of removing government subsidies and allowing the free market to govern trade in agricultural products.

Although freeing the international agricultural markets would make the worldwide trade in these commodities more rational, it could have adverse effect on the developing countries. All over the developing world, except for China, a period of profound economic change is underway. Called structural adjustment, this is in its broadest sense a transition from socialism to free market economies. It is having, and will continue to have great implications for food production and consumption. Liberalisation of economic policies in India has started yielding good results.

Environmental Accounting

As economies around the world undergo these rapid changes, their impacts on the environment are not always considered. One reason is that the politicians, and the senior civil servants or the bureaucrats who advise them, are not aware of the real costs of soil erosion, depletion of nutrients, and pollution. Those who are aware of these factors do not know how to integrate them into a policy framework; or they may simply ignore them in responding to them for the moment.

Putting a value tag on an environment such as a piece of land or a body of water is difficult. An approach to doing this is to survey the resource (such as land) and quantify its possible economic uses (such as forest, crop growth, fallow etc.). Then some measure of ecological stress is developed (for example, nutrient depletion leads to a transition from crop growth to fallow) and an economic value is put on this change. This must include not only the reduced yields from the resource, but also the costs of clean-up (for pollution) and regulation. If this process is done for a region or a country, a monetary value can be put on the rate of resource depletion leading to lowered economic value.

Some countries, notably Norway and France, have attempted to make annual environmental balance sheets, reporting "stocks" of renewable resources (such as farmland), non-renewable resources (such as minerals) and cyclical resources (such as air and water). These are then used as a guide to government policies in using these resources. These approaches require not only trained people to conduct the inventories of the natural environment and analyse them, but policy makers educated in these fields.

Food Security

When policymakers use the term 'food security', they are usually referring to grain stocks in excess of demand, left in storage against bad crop years. This has been a concern of planners, since agriculture began. But there is another aspect of food security, at the level of the individual household. True food security means that each person has a reliable, available, and affordable food supply. The United Nations sponsored World Commission on Environment and Development, set up to prepare for the U.N. conference on these subjects in Rio de Janerio in 1992, used the term equity as a criterion for distributing the world's resources, including food, and defined the term in two ways.

1. Equity for people living now who do not have equal access to natural resources or to social and economic goods.
2. Equity for the human generations to come, whose interests are not represented by current economic analyses or by market forces that do not take the future into account.

But, at the Rio de Janeiro conference, the Third World countries demanded that in the name of equity, their people should share agricultural resources, such as genetic strains of crop plants. Because much of the biological diversity is in the poorer countries, they could derive monetary benefit by ownership. Although most countries agreed to this principle, the United States initially refused to sign the treaty, under pressure from some industries that felt that exploration would be made less economically viable.

ALTERNATE AGRICULTURE

Plant growth and crop production are complex processes that depend on many interactions between organisms. The trend of modern agriculture has been:

- To substitute monoculture and continuous culture for crop rotation and diversified agriculture
- To use genetically more uniform plants, that have a narrow genetic base
- To use inorganic fertilizers, rather than the more difficult-to-use organic manures, in combination with green manures
- To use herbicides and pesticides to combat weeds and pests, rather than more complex biological control mechanisms
- To make bigger fields by eliminating all vegetation between them.

Recently, there has been a resurgence of interest in alternative agriculture, organic farming and sustainable agriculture. In 1989, the Board on Agriculture of the National Research Council of the United States lent credence to this movement with the publication of a major study called "Alternative Agriculture". Alternative agriculture is not a single system of organic, low input, regenerative, or sustainable farming. All these systems share an emphasis on management practices and on biological relationships between organisms.

Alternative agriculture recognizes that a piece of land on which crop plants are grown is first and foremost an ecosystem, and not a factory. An ecosystem has many interacting organisms that must remain in balance. Many natural processes occur in such an ecosystem and farmers should take advantage of these natural processes, rather than try to circumvent them or destroy them with chemicals. Alternative agriculture rejects certain practices (such as heavy use of inorganic fertilizers), but most important are the practices it favours:

1. Tillage that minimizes soil erosion even if it is more expensive
2. Reliance on animal manures and green manures with minimal input of inorganic fertilizers
3. Integrated pest management for pest control
4. Management systems, such as crop rotations, that help control weeds and disease organisms
5. These techniques are not new of course, although some of them have been refined in recent years.

Organic farming is defined as production system which avoids or largely excludes the use of synthetically compounded fertilizers, pesticides, growth regulators, and livestock feed additives. To the maximum extent feasible, organic farming systems rely upon crop rotations, crop residues, animal manures, legumes, green manures, off-farm organic wastes, mechanical cultivation mineral-bearing rocks, and aspects of biological pest control to maintain soil productivity and to supply plant nutrients, and to control insects, weeds and other pests (USDA, 1980).

In my view the following definition of organic farming seems to be more appropriate as it considers the important aspects like sustainability of natural resources and environment. **Organic farming** is a production system which favours maximum use of organic material (crop residues, animal excreta, legumes, on and off farm organic wastes, growth regulators, biopesticides etc.) and discourages use of synthetically produced agro-inputs, for maintaining soil productivity and fertility and pest management under conditions of sustainable natural resources and healthy environment.

Sustainable Agriculture

Sustainable agriculture, emphasizes the conservation of its own resources. For a farm to be sustainable, it must produce adequate amounts of high-quality foods, be environmentally safe, and where appropriate, be profitable. Sustainable farms minimize their purchased inputs (fertilizers, energy and equipment) and rely, as much as possible on the renewable resources of the farm itself. This is especially important in the 90 per cent of farms that exist in the third world, where these inputs are often not available or affordable.

Sustainable farming uses some form of integrated pest management for pest control, and this can include the use of chemical pesticides that are not used by organic farmers. Thus, sustainable agriculture does not mean a return to the farming methods of the late 1800's. Rather, it combines traditional techniques that stress conservation with modern technologies, such as improved seed, modern equipment for low-tillage practices, integrated pest management that relies heavily on biological control principles, and weed control that depends on crop rotations. Sustainable farms try to use wind or solar energy instead of purchased energy, and use organic animal manure and nitrogen-fixing legumes as green manure to maintain soil fertility, as much as possible, thereby minimising the need to purchase inputs from outside the farm. The use of genetically engineered crop strains is certainly not excluded by sustainable farming. The emphasis is on maintaining the environment, not on rules about what can or cannot be done. Profits from sustainable farms can exceed those of conventional farms.

Innovative farmers have developed many alternative farming methods and systems. These systems consist of a wide variety of integrated practices and methods suited to the specific needs, limitations, resource bases and economic conditions of different farms. To make wider adoption possible, however, farmers need to receive information and technical assistance in developing new management skills.

Sustainable Livelihood

Scientists, the world over, have developed new technologies for application in the field. Over the years they have come to realize that these technologies must be integrated into existing farming systems, and have been going out into the fields to try to understand the problems and to see how the solutions are working. The large-scale adoption of technology-based agriculture can lead to a loss of knowledge of traditional farming methods, which were developed over many centuries and are adapted to the physical, biological and social environments of the people. The knowledge may be useful in devising strategies for sustaining agriculture in the developing regions. Traditional agricultural methods may be applicable for designing sustainable systems in the developed countries. In other words, the rich and sophisticated can learn a lot from the poor and unsophisticated.

During the past one decade there have been important changes in the thinking of agricultural scientists and policy makers concerned with agricultural research in developing countries. It is clear that the type of research that led to the Green Revolution, has benefited the resource-rich farmers, whereas resource-poor families have been left behind.

It is increasingly recognized that who produces food and where it is produced may actually be as or more important than how much is produced. If resource-rich farmers produce food that subsistence farmers cannot purchase, there is no improvement in the food situation. Population projections show that many rural areas with fragile ecologies will have to continue supporting large populations.

Food can be produced in a more sustainable way that protects the resource base. The techniques are known and available and used by some farmers. Although more research is needed, these techniques could find more widespread use given the proper incentives. Currently, the incentives (price supports, tax policies and export subsidies) favour industrial agriculture. Society needs to replace them with incentives that favour sustainable agriculture. Because sustainable agriculture uses less inputs, this may cause some dislocation in the industrial sector and will be strongly resisted. In the Third World, much research on indigenous farming methods is needed, and the top-down approach to research needs to be replaced and/or supplemented by on-farm research. According to Dr. M.S. Swaminathan, "The greening of agriculture requires the greening of both technology and public policy. Producing more food and agricultural commodities from less land, water and energy is a task that will call for the integration of the best in modern technology, with the ecological strengths of traditional farming practices.

Organic Farming- Principles and Practices

The more widely used term organic farming describes two major aspects of alternative agriculture- (1) the substitution of manures and other organic matter for inorganic fertilizers, and (2) the use of biological pest control instead of chemical pest control.

Until well into the twentieth century, organic farming was not alternative agriculture, it was the worldwide way of life. It still is in many of the poorer regions of the Third World, where farmers cannot afford the technological inputs of modern agriculture. In North America, the Mennonites and Amish have long practiced organic farming out of tradition and choice. Studies of their farms show somewhat lower yields (5-15 per cent) than conventional farming in nearby areas. But their net return on investment is usually higher, because they consume less inputs, and when environmental costs are taken into account, the organic alternative is clearly superior.

Organic farmers emphasize using only organic fertilizers for fertility maintenance and banning nearly all chemical pest control methods. However, gypsum, an inorganic calcium salt, and rock phosphate are allowed as fertilizer, and a number of natural sprays are also allowed. Among these are sulphur dust, extracts of toxic plants, insecticidal and herbicidal soaps, and virus sprays. Organic farmers use antibiotics derived from fermentation (such as streptomycin), as long as the fungi that produce them are not allowed, but crops improved by traditional breeding are permitted.

In many respects, organic farming is a way of life as much as it is a method of farming. The profitability of organic farms depends on the higher prices that their products command in the market place. To stimulate organic farming, some governments have passed laws that create a demand for organic foods. For example, in some states of USA, poor people who receive food aid get coupons only redeemable at organic markets. Cities have created farmer's markets, where organic producers can sell their goods.

Organic agriculture is not based exclusively on short-term economics, but also considers ecological concepts. It utilizes appropriate technology and appropriate traditional farming methods. This form of farming can also be called sustainable agriculture. The principals of this method are:

- Organize the production of crops and livestock and the management of farm resources so that they harmonize rather than conflict with natural systems.
- Use and development of appropriate technologies based upon an understanding of biological systems.
- Achieve and maintain soil fertility for optimum production by relying primarily on renewable resources.
- Use diversification to pursue optimum production.
- Aim for optimum nutritional value of staple food.
- Use decentralized structures for processing, distributing and marketing of products.
- Strive for equitable relationships between those who work and live on the land.
- Create a system which is aesthetically pleasing for those working in this system and for those viewing it from the outside, *e.g.*, it should enhance rather than scare the landscape of which it forms a part.
- Maintain and preserve wildlife and their habitats.

Several practices in organic agriculture are aimed at minimizing the downward movement of nutrients in the soil profile:

1. The rotation with deep-rooted crops,
2. The avoidance of high-solubility nutrient sources,
3. The avoidance of mouldboard ploughing in favour of chisel ploughing,

4. The insertion of nutrients into the rotation onto a sod crop, if possible, to maximise uptake, and
5. The seasonal use of cover crops in and around the major cash crops.

We should observe a difference in nutrient profile with depth between organic and conventional systems if the assumption is correct and if any of the above practices have a major effect in vertical nutrient movement.

Approaching organic agriculture from the vantage point of a scientist trained in conventional farming techniques, Harwood (1984) outlines the following assumptions:

- Organic agriculture is a reversion to the agriculture of the 1940's.
- Organic agriculture is merely conventional agriculture without the use of chemicals. This assumption often leads to the use of a no-treatment check plot in a conventional experiment as an organic treatment.
- Organic agriculture is relevant only to backyard gardens.
- Organic agriculture is based on the assumption of a difference between chemicals produced by biological organisms and the same chemicals which have been synthesized in an industrial process.

Status of Organic Farming

Organic Farming in USA

The American Society of Agronomy (ASA) and the Soil Science Society of America (SSSA) are very concerned about developing alternative farming systems. For this reason, they held a symposium on organic farming in 1981 at Atlanta (USA). The objectives were to describe the current state-of-the-art of organic farming, describe strengths and weaknesses of the system, determine research needs, and provide a basis for development of alternative farming methods. Equally important, the Societies wanted to focus the attention on organic and alternative farming systems and to draw public attention to some of the severe problems facing agriculture.

During this symposium, the underlying question was: "Can organic farming contribute to a more sustainable agriculture, *i.e.*, one that is less dependent on intensive tillage and high inputs of synthetic and manufactured chemicals, while maintaining or increasing productivity and protecting the environment?" The most probable answer is that it most definitely can, and that our crop production systems can best be served by combining principles of organic agriculture with those of conventional and conservation tillage agriculture. Before this can be effectively accomplished, we need to develop a better understanding and appreciation of the organic farming system. As Dr. Richard Harwood pointed out during this symposium, "Most have the view that organic farming is conventional farming without synthetic fertilizer and pesticides, which is not true; organic farming is a complete farming system of its own." Most certainly the same relationship applies to conventional farming vs. no-tillage farming. Thus, as we attempt to develop a sustainable farming system, we must view organic farming as a separate entity. Whether this cropping system be called organic, low energy, biological, ecological, bio or sustainable farming is academic. For our purposes it could be called a biological farming system and be defined as: A farming system that uses minimum tillage, minimum pesticide, and synthetic fertilizer inputs, manages crop residues, manures and other organic matter. It minimizes energy inputs as much as possible and maximises the use of crop rotations for high crop yields and improved pest control. Sanitation, crop rotations, proper spacing of crops that are common hosts to pests, resistant crop varieties, crop varieties that are competitive with weeds and timeliness of tillage and planting are all important components for successfully controlling pests in the cropping system. Achieving this type of a farming system for even a portion of our agriculture will require a strong integrated research effort. Many of the participants in this symposium stressed this point.

The development of a sustainable farming system will require long-term interdisciplinary research addressing crop rotations, reduced tillage, efficient resource use and conservation, reduced use of synthetic fertilizer inputs, new or improved crop varieties, pest control, and the inter-relations of these management objectives. Management techniques are needed for increasing soil organic matter levels and improving nutrient utilization. Initial theory for developing these techniques will require a mechanistic approach. We have not had to address these questions at this magnitude or level with current or previous farming systems. Now economics, dependence on foreign oil, severe erosion, and increasing input costs dictate these approaches to ensure a sustainable farming system and to preserve the soil productivity.

Some of the research conducted during the era before the use of synthetic and manufactured fertilizers and pesticides has application to today's organic farming. Since that time, much progress has been made in areas such as plant nutrition, pest control, and plant breeding, to name a few. But, this research has been conducted under high fertilizer and pesticide inputs usually in conjunction with intensive tillage and has little direct application to the goals and aims of organic agriculture. Also, many present-day crop production systems have evolved into mono–cultures or make limited use of crop rotations, particularly those that are sod crops. The knowledge and practices learned from these systems may have limited applicability for meeting the rapidly changing needs of future agriculture, particularly with regard to development of sustainable systems. Moreover, we must recognize that organic or biological farming is a unique farming system and that its future progress will depend on development of new technology, not on technology developed 30 to 40 years ago.

The paucity of information on organic or ecological farming has been blamed to lack of public and scientific acceptance and support of research on alternative agriculture. A contributing factor is that scientists have not been willing or able to address the problem because of our short-term accountability climate. Many of the answers needed will require a great deal of research before publication is possible. Long-term, in-depth studies have not been popular or supportable.

There is a pressing need for information and research in organic agriculture in a number of areas. For example, we have little or no information on how modern crop varieties perform under intensive rotations which include legumes and reduced mineral fertilizer additions in conjunction with reduced tillage or no-tillage. There are few data on the interactions among weeds, crop varieties, and fertilizer levels and the net effect on herbicide need and efficacy. Has the heavy use of synthetic and manufactured fertilizers increased weed or insect infestations in crops or plant diseases? We have little information on how much an organic farming system will reduce erosion, decrease nutrient losses, and maintain soil productivity. We do not know if cropping systems can be managed to more effectively mine nutrients such as P or K from the soil minerals.

The biology of soil fertility, the impact of synthetic and manufactured fertilizer inputs, nutrient relationships to tillage and crop residue management, and resultant interactions and effects on crop nutrition and soil organic matter require more study.

There is much that we do not know about the soil-plant nutrient relationships when crop production, soil and residue-management systems are changed. We do not completely understand the value of organic N vs. mineral N or the factors controlling symbiotic nitrogen fixation. The difference may be one of timing of N availability, the efficiency of use or some process we do not understand.

Nutrient use efficiency continues to be a major concern in agriculture. The average use efficiency of fertilizer N in many cropping systems is about 50 per cent. During the season of application, plant uptake of applied P fertilizer ranges from 5 to 25 per cent (Hauck and Koshino, 1971). Additional P uptake occurs in following crop seasons, but its use efficiency is still lower than we would like. Some research data make the P use efficiency picture even more puzzling.

There are many studies showing that mycorrhizae can aid P nutrition of some plants under certain conditions. Potentially, these microorganisms, if we can determine conditions favouring their development, can be used to ease pressures on our non-renewable P resources. The positive effect of soil organic matter on soil properties and nutrient recycling is undisputed and yet we do not understand the dynamics of the system that control soil organic matter levels. Maximum use efficiency and recycling of soil nutrients and maintaining or increasing soil organic matter levels is essential for a sustainable agriculture.

There is no question current conventional farming systems can be made more energy efficient, more protective of the soil, and less prone to environmental pollution. Fertilizer costs in previous years have been low relative to economic returns resulting in application rates out on the plateau of the crop response curve. Fertilizer rates adjusted to the peak of the response curve or lower, coupled with soil testing, will reduce waste, pollution, and costs, and will not likely reduce crop yields appreciably. There is evidence indicating that high fertilizer N applications are increasing NO_3^--N in some aquifers (Schepers and Mielke, 1982). Also, many farmers make too many tillage trips across their fields which waste energy, destroy soil organic matter, and increase the erosion hazard. Over application of pesticides or poor timing of application result in waste and possible contamination of water supplies (Wehtje *et al.*, 1982). Training for proper pesticide usage is very important. Over use and improper management of irrigation water are wastes of a resource and can also result in pollution.

There is another point that should be addressed. Traditionally, many people have thought of organic farms as small operations. Yet, in the USA, there are organic farms as large as 600 ha (USDA, 1980; Lockeretz *et al.*, 1983). In view of the severe weed and pest problems we often encounter with farming systems, it does not seem possible that pests can be controlled on large farms without chemical pesticides, but apparently some organic farmers are doing this successfully. There are several reasons why organic farmers are controlling pests. Successful organic farmers stress the need for crop rotations and timing of operations such as tillage (normally they practice shallow tillage, which affects the ecology of the system) and planting time as being critical. In many cases, they do use pesticides, but sparingly. Crop rotations and green manures reduce pest problems. Biological control of pests has also been used successfully in many cases. The areas of entomology, plant pathology, and weed control have made some impressive advances in this area; but much remains to be done in all areas of biological control of crop pests. As these technologies emerge, we must adopt them and use them effectively.

Another aspect of organic farming is technology transfer to developing countries. These countries need better technology to produce food for themselves. It would appear that organic agriculture could better serve their needs than could conventional energy-intensive agriculture. However, educational system in USA generally does not provide the training or expertise to grow crops using low-energy inputs and other practices that are characteristic of organic farming systems.

Organic Farming in Europe

Very similar to the situation in the USA there is an increasing interest in organic or biological farming practices in Europe. On a world-wide basis, many of the indigenous farming practices could be called organic farming. However, the modern type of scientifically based biological farming methods are practiced mainly in Europe and North America. Organic agriculture, based on the work of Howard (1943) and Balfour (1978), is predominant in the English speaking world. The biodynamic movement, based on anthroposophical principals and founded by Steiner (1924), can be found throughout the world, but it is very strong in Europe. The biological movement, founded by Hans Mueller in Switzerland in the 1940's and based on the findings of Rusch (1968) from Germany is mainly found in German-speaking countries, but is spreading into Scandinavia and other countries as well. In France the group of Lemaire-Boucher founded in the early 1960's is growing rapidly (Saint Heinis, 1972).

In Germany, the Bio-Dynamic Research Institute at Darmstadt, and in Sweden, the Scandinavian Research Circle at Jarna have been active in biological research. Publically-funded biological farming activities are carried out in Holland and Germany and to a lesser extent in Switzerland, Austria, and Sweden. In Holland and Bavaria, there is an increase in the number of official extension agents trained for advising biological farmers. In Austria, the Government has appointed one agricultural advisor who is responsible for coordinating all the activities in the field of biological agriculture.

Most of the examples of highly industrialized agriculture can be seen in Europe and North America, where highly-developed biological farms occupy from 0 to 1 per cent of the total farms and are on the increase. Studies by Meili (1979) and Gebhardt (1982) showed that in Switzerland, approximately 10 per cent of the conventional farmers were prepared to alter their farming system to a more biological approach. In Germany for the part of Hessia, this figure was 27 per cent.

Experiments on biological agriculture are under way in various countries. In Germany three government-owned farms have been converted to biological agricultural methods. The first Chair for a Professor of Biological Agriculture was created at the University of Kassel, Germany. In Switzerland most of the research activities are funded by the Swiss Research Institute of Biological Husbandry, a privately financed institution. Lectures on biological farming are given in various universities throughout Europe and are partly within the compulsory programme of these universities. In some cases, these lectures were established due to pressure from student groups. In general, environmental groups favour and support biological farming in Europe. The 1982 International Scientific Conference of the International Federation of Organic Agriculture Movements (IFOAM) was held at the Massachusetts Institute of Technology, Cambridge, Mass. The previous international scientific conferences on biological agriculture were held in Switzerland, Canada, and Belgium.

Traditionally, small farms had combined several kinds of agricultural land use; *e.g.*, arable farming and animal husbandry or horticulture. This mixed system of farming disappeared to make room for a more specialized type. This change has led to environmental problems in some areas with the manure from a large number of animals and where there is an intensive use of mineral fertilizers and pesticides. In countries like Belgium, Holland and England this situation is pronounced. Holland has the highest use of chemical fertilizers in Europe (FAI, 1994) and simultaneously the highest production of manure due to imported feed.

The European biological farming systems have a number of common characteristics which are found to various degrees in biological agriculture systems in other parts of the world and even to some extent in more intensive agriculture farming systems. Some of these elements are:

- Balanced crop rotations with legumes and catch crops are used.

- Green cover of the soil is maintained as much as possible with green manures and under sowing.
- Animals are preferably included in the farming operation.
- Shallow ploughing (approximately 12 to 15 cm), is practised with deep loosening of the soil to a depth of 30 cm.
- The purpose of fertilizing is to activate soil life and improve soil structure (no use of synthetic mineral-N)
- Weed control is conducted through the cropping system, tillage and mechanical as well as thermal methods, *e.g.*, flame gun, etc. (no use of herbicides).
- Pest control is conducted through careful selection of crop varieties that are best adapted to local conditions, cropping systems, and tillage. Various plant extracts as preventatives are used. Biological and biotechnical methods are also used. The direct application of certain compounds is allowed (*e.g.*, Rotenone, Cu, S, etc.) (no use of synthetic pesticides).

From this it becomes obvious that the crop rotation plays an important part not only for weed control and soil improvement, but also for the supply of N. The other main elements are supplied through basic slag, rock phosphates, muriate of potash etc.

Organic Farming in India

Green Revolution in India was mainly realised with the introduction of high yielding varieties of various crops and by following intensive cultivation practices with the use of fertilizers, pesticides and other inputs. The fertilizer consumption in the country accordingly has increased from 66,000 tonnes in 1951-52 to about 12.16 million tonnes in 1992-93. In other terms, fertilizer consumption which was only 0.5 kg/ha in 1951-52, increased to 67 kg during 1992-93. Same was the situation with pesticides. All these agro chemicals are produced from non-renewable fossil fuel. In future, diminishing availability of these inputs may not only cost heavily on our foreign exchange but may also limit agricultural production.

With the increase in the country's own population, compulsion would be not only to stabilise agricultural production but to increase it further in sustainable manner. The scientists have realised that the green revolution with high input use has reached a plateau and is now sustained with diminishing return and falling dividend. The intensive use of inputs has not only polluted the soil, water and the environment causing their slow degradation but also affected the human beings. It is because of these the health conscious countries like Germany, Japan, USA, Netherlands etc. and the affluent domestic population are ready to pay a premium for food raised without the use of chemicals. Thus, a natural balance need to be maintained for survival and well being of the human beings and plant and animal kingdom. The obvious choice for that would be adaptation of organic farming practices without compromising agricultural production.

Reports of deleterious effects arising with the use of agrochemicals was being observed by the Government of India for quite sometime. This situation, coupled with the degradation of cultivable land and increasing agricultural pollution has created an unhealthy situation in the country. In order to balance this situation, it was felt to have a viable programme which may not only check soil degradation, underground water pollution, high residues of hazardous chemicals in food, but may sustain the food production and improve the soil health without effecting crop production. Organic farming, which aims at cultivating the land and raising crops in such a way as to keep the soil alive and in good health may be an alternative to the present system of farming solely depending on chemicals.

To study the viability of organic farming in India, a Technical Team constituted by the Ministry of Agriculture visited several research centres and farmers practising organic farming during 1993 and made the following observations:

The country at present is not in a position to completely eliminate the use of chemicals especially fertilizers. However, it would not be difficult and unrealistic to phase out the use of these chemicals systematically. For this, on one hand, the doses of fertilizer need to be gradually reduced and be balanced by increasing the use of optimum quantity of organic manures and biofertilisers. Most of the State Agricultural Universities are doing very little work in minimising the use of chemicals and increasing the use of organic manure. Commercial sale of organic manure is uncontrolled and needs proper quality control standards so that the poor cultivator may not be cheated on one hand and on the other the ecosystem may not be disturbed. Some of the States have not incorporated the use of farm yard manure even in their package of practises. Farmers are aware of the hazards of pesticide use and want to minimise its use but absence of any knowledge about alternate uses forces them to continue use of pesticides. Neem and other materials of plant origin for pesticide use are not readily available. For the control of diseases and pests, the team was fully convinced that Integrated Pest Management is the answer for checking indiscriminate use of pesticides. The plant extracts have been used by various farmers for controlling insects and diseases with sufficiently high efficiency and various combinations may be tried.

The Extension Staff of many States is also unaware of the beneficial use of organic manure and deleterious effect of agrochemicals. As such, their efforts in propagating organic farming is negligible.

Components of Organic Farming Systems Research

Meaningful research component in organic farming systems has been difficult to design, primarily because of the nature of organic philosophy. There is no shortage of meaningful component research which could be done on methods of increasing N2 fixation in legumes on component crops and on weed and insect control measures. This research would touch only marginally on the basic questions inherent in organic philosophy.

There are assumptions throughout the organic literature of differences between organic and conventional systems with respect to their effects on soil physical properties, soil insect fauna, nutrient flow within the soil, crop health, and nutritional value of the harvested crop. One finds many principles commonly held by organic farmers such as feed the soil and not the crop, and the need to rotate with deep rooted crops to bring nutrients from deep in the soil profile. There are other perplexing observations such as the decline in yields during the process of conversion from conventional to organic practices. These assumptions, principles and observations cover virtually every aspect of crop and animal management in organic systems. Some are readily understandable, but many deal with physical and biological interactions which have not been researched and are little understood.

Crop and Soil Management

Organic farming systems encourage the use of rotations and manures to maintain soil fertility. Carefully managed soils with a high proportion of humus offer essential advantages with respect to water retention, ion exchange, soil erosion, and animal life in the soil. A high proportion of humus in the soil gives uniform distribution of nutrients and also promotes plant hygiene.

Green manuring and intercropping of legumes is another important aspect for biological farming systems not only in regard to weed control, but also in reducing the leaching of nutrients and in reducing soil erosion. A green cover throughout most of the year is one of the main goals of such farming methods. Depending on the green manure mixture or the legumes used for under sowing, there may be an increase in soil organic matter and soil N as well as in other nutrients.

On Farm Waste Recycling

Increasing prices of chemical fertilizers have enabled organic wastes to regain an important role in the fertilizer practices on the farm. Good manure management means improved fertilizer value of manure and slurry and less nutrient losses. Composting of all organic waste in general, and of farm yard manure or feed-lot manure in particular, is important in organic farming.

Non-Chemical Weed Management

Weed management is one of the main concerns in organic agriculture. Generally, all aspects of arable crop production play an important role in a systems approach to problems. The elements to consider in preventing weed problems are crop rotation, green manuring, manure management and tillage. In addition, mechanical and thermal methods are used (flame guns) in the pre-emerging system. Mulching on a large scale by using manure spreaders may also be useful in weed control.

Domestic and Industrial Waste Recycling

Sewage and sludge use for crop production can form an important component of organic farming if treatment and application methods are improved further.

Energy Use

In the USA, Lockeretz (1978) has demonstrated that the energy requirement for production measured per dollar of produce for organic farms is only one-third of what it is for their conventional counterparts. Because mineral N-fertilizers and pesticides (require large energy inputs for their production) are not used by biological farmers, the comparison of total energy input per hectare with total energy output favours biological farming systems (Lunzer, 1980). However, the reduction in energy use on biological farms is similar for all studies.

Food Quality

Food quality is one of the main issues, which concerns both scientists and consumers. One of the difficulties is that analyses of nutrients and various compounds are not sufficient for a well-founded assessment of the quality of agricultural produce. Nitrates in water and farm produce, desirable components, pesticide residues, keeping quality and physiological imbalances are some of the important aspects of food quality. For a fair evaluation of quality, it may be necessary to use many more analytical parameters than are commonly used, or even consider other aspects that go beyond a purely chemical interpretation. In this field a highly promising start has been made concerning the storage and processing ability of staple farm produce as well as in microbiological and physiological tests. The ideal approach would be to grasp the effect of a whole system on another system- a comprehensive, interdisciplinary approach. Such programmes should unite scientists in the fields of agriculture, food science, nutrition and medicine. However, such comprehensive programmes have not been realized so far in practice, although plans do exist.

Ecological Agriculture

The growing concern about environmental degradation, dwindling natural resources and urgency to meet the food needs of the increasing population are compelling farm scientists and policy makers to seriously examine alternatives to chemical agriculture. A sustainable agriculture backed-up by "green technologies" in an integrated farming system has been considered a promising and potential pathway. The twin problems confronting agricultural production are all pervasive erosion of natural resources such as land, water and biodiversity, fast declining soil fertility and use efficiency of inputs such as water, fertilizer and energy. Demographic pressures accelerate the first and agronomic deficiencies the second.

The gravity of the environmental degradation, arising from faulty practices, has set several experts in the field to focus attention on ecologically sound, viable and sustainable farming systems. Though there has been a series of scientific and policy conferences on this issue, the sustainable farming methods are yet to get the approval of most agricultural scientists.

There are no serious attempts to develop a package of practices for large-scale adoption of sustainable agriculture in India. Only a few enterprising farmers and private institutions stand testimony to the utility of safe and highly productive natural farming methods.

The adverse effect of pesticides on the environment has been well documented, and their residues in the food chain have endangered the whole life sustaining systems in many regions. Chemical fertilizers have also jeopardised the environment through nitrate poisoning and exterminating the beneficial soil microflora and microfauna by adversely altering the chemical and physical structures of the soil.

Objectives of Ecological Farming

Bringing to match the crop, soil and climate of a region the ecology and farming and gaining from the economy and efficiency of inputs are the objectives of ecofarming. According to Sankaram (1996), this technology may be grouped into four categories.

1. Those that reduce the environmental burden of green house gases (CO_2 and CH_4), global warming, ozone depletion. All attention to promote renewable sources of energy (draught animal power), electrical energy from garbage disposal and biogas from organic wastes.

2. Those that reduce the demand on land, water and biodiversity without adverse effect on agricultural production and nutritive value of food. Nurse the soil back to health, change cropping patterns to maximise ecological productive efficiency, improve, water use efficiency- through conjunctive use of rain, tank, underground, well and river waters; reduce conveyance losses; phase out subsidies.
3. Those that continue to improve crop productivity, under shrinking land resources. (genetic and agronomic) hybrid vigour, gene pyramiding; multiple cropping patterns, integrated nutrient management and integrated pest management.
4. Reduce hunger and poverty, adopt, cost effective farming to bring equity of food price and wage, encourage job promoted growth right in the village to arrest migration as ecological refugees.

Ecological farming systems

The shift from chemical to ecological agriculture should, however, be gradual. A sudden switch-over can spell disaster and discourage farmers from taking to this course. At least two to three seasons will be needed for the transition, and during the interim years the farmers could build up a sufficient organic base to fertilise the fields and improve the productivity of the soil. Once a state of ecological balance has been attained and there are a good number of beneficial organisms to check the explosion of pests and pathogens, crop yields of a high order can be ensured. The cost of cultivation can be brought down substantially as many farm-grown inputs can be integrated efficiently in the farming system.

Ecofarming can be sustainable and profitable too. It is a self-reliant method and has to be an integrated system. Since it is a knowledge-intensive practice, one has to keep pace with the dynamics of nature to increase the biological productivity of the soil.

Since ecofarming uses several farm grown inputs, and is less dependent on market purchased inputs, it is economically attractive to the growers. As reported by Venkataramani (1995), case studies show that when a chemical farm incurred about

Rs. 11,250 towards the cost of cultivation per hectare of rice, an organic farm spent Rs. 10,590. Alternatively, if the biofertilisers and organic nutrient supplements such as neem cake are subsidised to extent the same monetary advantage as chemical fertilizers, the cost of ecological farming will come down significantly. It has been observed that in chemical farming, fertilizers and plant protection chemicals accounted for nearly 30 per cent of the total cost. The cost of market purchased inputs in ecological farming was far less.

In both the cases, the yield remained almost the same at 5,625 kg paddy per hectare. The straw yield was eight tonnes. The net returns from the ecological farming system at the current cost of Rs. 3.34 per kg paddy is Rs. 8,197.50. In chemical farming, the net profit is Rs. 7,500. If one gets a premium price for the poison-free, organically grown rice, the economic returns from the ecological returns from the ecological farming system will be highly encouraging.

While the modern system of farming expended nearly 31,000 MJ of energy, the ecological farming method accounted for only 23,400 MJ. High energy inputs such as fertilizer, plant protection chemicals, herbicides and high speed diesel are substituted with biological inputs and low-energy human and cattle power. The output/input ratio in the case of modern farming is 3.76, while in the case of ecological farming it is more than 4.95.

Ecological farming is more labour intensive, and several of its components such as vermicompositing, vermiwash, and generation of botanical pesticides and rearing beneficial insects such as Trichogramma parasites, ladybird beetles, and spiders for release in the field as biocontrol agents offer scope for unemployed and landless youth.

Group and community endeavour holds the key for successful implementation of ecological farming over larger areas. Community participation in pest control and rodent eradication will be rewarding. Even in the case of developing biofertilisers such as blue green algae and *Azolla,* group action will pay rich dividends.

Integrated Intensive Farming System (IIFS)

According to Prof. M. S. Swaminathan, the Integrated Intensive Farming System (IIFS) methodology provides the pathway to achieving an evergreen revolution in agriculture. The IIFS involves agricultural intensification, diversification and value-addition. It helps improve physical and economic access to food, there by fostering sustainable food security at the level of each individual in a house hold. He is of the view that emphasis on food security at the individual level is important in view of the growing feminisation of poverty, according to him.

IIFS involves intensive use of farm resources. To be ecologically sustainable, such intensification should be based on techniques which are knowledge-intensive rather than capital-intensive, and which replace, to the extent possible, market-purchased chemical inputs with farm-grown biological inputs (Venkataramani, 1996).

Such a shift in the nature of inputs used is brought about through integrated farming involving animal husbandry, fishery and agroforestry. Such integrated farming practices which constitute the second element of the IIFS strategy provide scope for organic recycling. The third element of the strategy is value-addition to every part of the plant and animal biomass through the establishment of biorefineries. On-farm and off-farm employment can then be linked in a symbiotic manner. In the case of farm women, who are invariably over worked because of their multiple roles in a household, IIFS aims to reduce the number of hours of work and add economic value to each hour of their work, according to Prof. Swaminathan.

With an ever-expanding urban population, the demand for a wide range of processed and pre-cooked foods will expand, as urbanisation leads to diversification of foods habits. The IIFS with its emphasis on crop-live-stock-fish integration responds to this challenge. Also, the emphasis placed on post-harvest technology in IIFS helps both producers and consumers.

To summarise, IIFS involves a pro-nature, pro-employment, pro-women and pro-poor orientation to technology and dissemination. It leads to labour diversification and not displacement. It leads to resource-based, in contrast to the widely prevalent commodity-based agricultural development planning.

Prof. Swaminathan has prescribed **Seven Pillars of IIFS** for successful implementation of such systems. Soil health care is fundamental to sustainable intensification. In addition, vermiculture constitutes an essential component of IIFS. IFFS farmers should maintain a **soil health card** to monitor the impact of farming systems on the physical, chemical and microbiological components of the soil.

Water harvesting and management is the next crucial component. Crop and pest management including integrated nutrient supply and integrated pest management (IPM) systems form an important component as well. Energy management, post-harvest management, choice of crops and other components of farming system and information technology, skill, organization and management empowerment constitute the rest. This will help ensure economic viability, environment of sustainability and social and gender equity in IIFS villages.

It should be emphasised that IIFS will succeed only if it is a human-centred rather than a mere technology-driven programme. Its essence is the symbiotic partnership between farming families and their natural resource endowments of land, water forests, flora, fauna and sunlight.

Prof. Swaminathan has stated that without appropriate public policy support in areas such as land reform, security of tenure, rural infrastructure. input and output pricing and marketing, small farmers will find it difficult to adopt IIFS. The eco-technologies and public policy measures needed to make IIFS a mass movement should receive concurrent attention.

Ecological agriculture that effectively combines use efficiency of inputs and economic yield maximisation will begin to prevent the abuse of natural resources (Sankaram, 1996). On its progress depends the achievement of food security and sustainability of farming systems. Ecological agriculture will reactivate the fatigued green revolution to move forward through promotion of soil health, package of agronomic practices with each component a blend of tradition and modernity, and a policy to value each input based on its intrinsic worth free from subsidies. There is the need to recapture local knowledge about framing systems. A synthesis of traditional wisdom and ecological prudence of tropical agriculture with modern advances in technology will help higher use efficiency of inputs and sustainable crop yield maximisation. This is symbolic of **low input sustainable agriculture of (LISA)** and **sustainable agricultural research and education programme (SAREP)** widely promoted in the United States of America.

Agricultural research has to be directed towards a better understanding of agro-ecosystems and how to maintain them, rather than towards the destruction or replacement of biological systems with more energy consuming and industrialized methods. Concern for alternative solutions should be a goal of all research. The development of appropriate technologies is another important area of research.

CHAPTER 2

Sustainable Agricultural Systems: The World Scenario

The agricultural progress witnessed the world over in the past four decades has been quite impressive. This progress is very much linked with the demand of the increasing population especially in less industrialized and third world nations. However, the changes associated with the increase in food production seem to be alarming and have drawn attention of the governments, planners and the scientists. Soil degradation, extinction of biospecies, environmental pollution, global warming and several other factors which threaten the very existence of life on the planet have raised doubts about the pattern of development followed in different countries.

Thus, there is an urgent need to keep up the pace of the present development and ensure for prospects of development for future generations. The developmental activities related to all aspects of human life should be made sustainable over a period of time.

Swaminathan (1992) suggests that sustainability has ecological, economic, social and cultural dimensions. This is also true with agricultural sustainability. The demand of the increasing population can be met by adopting a sustainable approach towards biological productivity.

SUSTAINABLE AGRICULTURE

Sustaining the productivity of the earth's natural resources will be the most important issue in the twenty first century. There is an urgent need to develop more land use systems which shall address to requirement of food, fodder, fibre and fuel by the future generations.

The term sustainable agriculture is relatively new to many people. The existing systems emphasize high yields which are dependent on high input use. Alternate farming systems range from systems with only slightly reduced use of inputs through better soil testing facilities, integrated pest management, and capital inputs to those which seek to minimise their use through suitable crop rotations, integration of livestock with crops, mechanical and biological weed control and with less investments on infrastructure and equipment.

Sustainable agriculture is a loosely-defined term that comprises a range of strategies for addressing many of the problems that affect agriculture world wide. These problems include the loss of soil productivity from excessive erosion, pollution of surface and ground water from pesticides and fertilizers, diminishing supplies of non-renewable resources and depressed farm income because of low commodity prices and high production costs. The concept of sustainability involves economic, environmental as well as social aspects. A farming system using recommended levels of technical inputs may be economically sustainable at present but if it contributes to pollution of any kind, it is not sustainable over a long term. Also, if a farming system does not address to the demand of the society, it is not considered to be sustainable. The ultimate aim of sustainable agriculture is to develop farming systems that are productive and profitable, conserve the natural resource base, protect the environment and enhance health and safety over the long term. How to achieve this goal will depend on creative and innovative conservation and production practices that provide farmers with economically viable and environmentally sound alternatives in developing their farming systems.

Need for Sustainable Agricultural Systems

Alternate or sustainable agriculture is not a single system of farming practices. It includes a spectrum of farming systems, ranging from organic systems that attempt to eliminate the use of synthetic chemical inputs to those involving the proper use of pesticides to control specific pests or diseases. The ultimate objective is to maximise the benefits from the existing agricultural assets and minimise the threats to the environment from the current practices of intensive agriculture.

In recent years, agriculture has taken on new importance world-wide as the means to revitalize and expand the economies of many countries. Research and education in the field of agricultural sciences have had a high priority in these countries and, as a result, there has emerged a global network of scientists who are engaged in developing new technologies to meet the production demands under different agroclimatic conditions.

The exchange of scientific information is of vital concern to all countries if we are to address both economic and environmental issues in agriculture on a global basis. The feeling of concern is quite intensive considering the fact that the growing world population will need to use more and more of earth's natural resources. The global population is presently estimated to be 6.0 billion people. By the turn of the century, the population is expected to rise by another billion and by 2020, it is likely to exceed 8 billion. In order to meet the requirement of the growing population in respect of food and fibre and feed for the livestock, suitable sustainable agricultural systems need to be developed.

Sustainable development has quite suddenly become a global issue. The concept of sustainability has emerged from the synthesis of different perspectives, and hence it is not likely that commonly accepted definitions will emerge in near future. However, all proponents of sustainable agricultural development agree that efforts may be made to develop the land use systems that are economically viable in the short run yet not environmentally degrading in the long run. In short, the essence of sustainability is the maintenance of natural resource productivity (Hart and Sands, 1992).

The concept of sustainable land use systems may be one way to integrate and make operational these different perspectives. Sustainable land use systems can be conceptualized as farm systems that interact with their biophysical and socioeconomic environment in such a way that they are both economically viable in short-run and ecologically sustainable in the long run. One way of identifying potential strategies or research approaches is to analyse and consider alternative technology that could enhance- (i) farm-by-biophysical interactions, (ii) farm component interactions, and (iii) farm-by-socioeconomic environment interactions.

Sustainable development is underlain by an ethical norm focusing on a new relationship of cooperation and co existence of humans with nature as compared to the prevailing conquest and exploitation of nature by humans (Batie, 1989). Many of the concepts and tools employed at present are suited to technically efficient resource development. However, this quality of agricultural systems is inappropriate for a sustainable perspective that embodies a sense of ecological connection between the action of humans and the consequences for them. Sustainability reflects an extended concern for intergenerational equity (Baite, 1989). It is not a quantity that can be measured and tested simply and that is why it is so difficult to establish a definition of sustainability.

STATUS OF SUSTAINABLE AGRICULTURAL SYSTEMS IN ASIA

India

The strides made by Indian Agriculture in the past four decades have been impressive. Food grain production has increased synchronizing with the growth rate of population. However, the main spurt in production has been in irrigated sector and that too mainly in wheat and rice crops. These are the crops which received the greatest stimulus of varietal improvement, shared major proportion of irrigation and fertilizers and benefited most from Governmental price support and procurement policies. In contrast to them, pulses, oilseeds and coarse cereals which are cultivated mainly in the rainfed sector remained rather deprived of these benefits and are pushed to marginal areas.

Seventy percent of the cultivated area in the country lies in the region of medium to low rainfall (1150 mm and below) and in most cases the rainfall is inadequate and uncertain with respect to crop-water requirements. Higher level of crop production and its stability can be obtained only through irrigations. Till 1951, when the first five year plan was launched, India had an irrigation potential of 22.6 mha. The eight plan target provides for the creation of additional 15 mha potential of which 13.5 mha would be the target for utilization in the terminal year. The expansion of irrigation resources brought about not only enhancement in production but also stability in production.

Nutrient management is the key to higher crop productions. Most of the growth in the food production during the Green Revolution period is attributed to the higher fertilizer use. The annual fertilizer consumption is expected to rise to about 20 million tonnes in India by the end of this century. This rise in fertilizer use is necessary because we foresee, (i) N deficiency will continue to be universal in Indian soil, (ii) deficiency of P will be next to the order of the extent, (iii) K will become limiting in high productive regions, (iv) in at least half of the Indian soils, crops would benefit from Zn treatment, and (v) S deficiency will obstruct the optimum productivity in vast majority of Indian soils. We do not rule out the emergence of additional nutrient disorders which may stand in the way of sustaining the envisaged growth rate in agricultural productivity (Abrol and Katyal, 1990). The situation, therefore, calls for an integrated nutrient management and supply system which (i) integrates the use of on-farm generated organic manures with inorganic fertilizers, and (ii) exploits the natural nutrients supply through bioengineering and mining of soil resources.

With the advent of the green revolution period, the new crop varieties and cropping sequences for intensive agriculture, brought to the fore front problems of pests which caused tremendous losses to various crops and their produce. Pesticide consumption has shown steady increase. From the mid fifties application has increased both in quantity and in coverage. Plant Protection coverage increased from 2.4 million hectares in 1956 to 80 million hectares in 1984 and the quantity of pesticides rose from 200 tonnes in 1955 to 72,000 tonnes (tech.) in 1987. Their influence in upsetting the ecological balance and polluting the environment is being felt increasingly. Safe use of pesticides should therefore, be the target.

Good quality seed is one of the basic requirements for augmenting crop production and productivity. The extent of area under High Yielding Varieties (HYV) is one of the important indicators of the spread of modern technology for increasing agricultural production. The scope of exploiting the genetic potential of plants for higher yields under different input intensities and under different situations is tremendous. Development of appropriate crop varieties is going to be the main component of sustainable agriculture.

The research strategy for sustainable agriculture should be an exercise in the development of such farming systems which meet the production objectives through most efficient utilization of inputs without impairing the quality of environments with which the system interacts. Use of land according to its capability, integrated use of purchased inputs and farm residues of production, harnessing maximum productive efficiency from the inputs and from the interacting environments and considering these objectives in long term perspectives are some of the important aspects of sustainable agriculture strategy.

The need is to integrate the components and evaluate the synthesized systems against the existing system of production. Long term monitoring of the improved systems with regard to the parameters of sustainability will be required. On-station research could be initiated but the on-farm testing of the developed technology will be required ultimately. According to Singh *et al.* (1992), the following may be the major course of research work in future on sustainable agriculture in India:

1. Synthesis of the sub-systems of sustainable agriculture through on–station research.
2. Synthesis of a sustainable farming system based on the location specific needs through on-farm research.
3. Development of research methods for evaluating the farming systems with regard to its biological, environmental and social efficiency.
4. Development of methodology for monitoring the improved systems over long period of time.

China

China is mainly dependent on its agricultural economy. The diversity of climate and the complexity of topography in addition to the long history in the agricultural production result in the extreme complexity and diversity of China's farming system and the system of agricultural production. Under the different conditions of climate, soil and socio-economic formation, there exist different cropping systems and agricultural systems.

China casts to 22 per cent of the world population with only 7 per cent of the total tilled land in the world; and this is the significant achievement of China's agriculture. However, with the increase in population pressures, resources, food products, and environment, the problem of sustainable development in China's agriculture appears to be very important and urgent. Since 1980s, the developments of ecological agricultural have carried out tests and studies at selected points in accordance with the local socio-economic situations and natural conditions in many regions of the country and also achieved relatively marked economic environmental and social benefits. The studies and technologies related with ecological agriculture are being taken seriously by the state departments and welcomed by the farmers.

Some of the ecological farming technologies such as ecological engineering technique of benign cycle, comprehensive development strategies in agriculture and the technique of cubic utilization have been popularized among the peasants. With the benign cycle technique, it is possible to achieve an organic link between the related components in agricultural production through an improvement of the interrelationships among the crops, domestic animals and soil. The trash of one component will become input feed for another component, thus raising the utilization ratio of substances, increasing the productivity of the agroecosystem and enriching the diversity of the system as well.

By the technique of cubic (or multi-story) is meant that utilization of space and time on one same piece of land to form a layer-by-layer combination in production so as to raise the productivity from the same plot and adjust or improve soil fertility. This technique seems to have especially practical value in China which has less tillable land and more population. For instance, fish breeding in paddy rice field, cubic utilization of orchards, interplanting of mushrooms in paddy rice field, duck and fish breeding in the irrigated woodland of pond cypress, multiple layered artificial cultivation of coenosium, multiple cultivation in rubber plantation, various forms of intercropping and interplanting techniques, etc., are all being emphasized by the people.

China's ecological agriculture has certain similarity with the organic agriculture in other countries. The studies of Chinese eco-farming will make contributions to the sustainable development of agriculture (Zhengfang and Youxia, 1992).

Japan

The terminology of sustainability is quite new to Japanese agriculture (Miyazaki and Kada, 1992). The main reason for the lack of the concept of sustainable development is the fact that rice-based Japanese agriculture has developed and continued for nearly two thousand years under the temperate, monsoon climate. Needless to say, rice is produced in paddy field, which necessarily contains a certain depth and amount of water during its growing period. By this nature, rice production has possessed by itself an ecologically and biologically sustainable system, which has enabled farmers to produce rice in paddy field year after year, for a long time.

However, such a sustainable system has been gradually changing since the early 1960s, when Japan has entered into a highly industrial economic development stage. The rice production system has also changed from being one of relatively low-input and low-output to the one of high-input and high-output. In other words, Japanese agriculture has rapidly increased its intensity of land use per unit of land area, which has caused such problems as deterioration of soil fertility, soil and ground water pollution, and other conservation problems. Recently, Japanese people are increasingly concerned about these issues, as global environment problems such as global warming and desertification have become a serious global issue.

In order to increase the food self-sufficiency of domestic production and to utilize paddy fields more efficiently, one of the most important agricultural policy goals of present Japan is how to reorganize cropping patterns of paddy fields under these changing circumstances.

For this, the Ministry of Agriculture had adopted and implemented a long-term project to remodel the system of land use toward one more ecologically adaptive and economically viable. One of the core plans is to reorganize the use of paddy such as wheat, barley, soybeans or forages, on a regional basis. To this end, it is encouraged, for example, to establish the so-called **manure bank** system in each locality, for the purpose of utilizing organic manure (substituted for chemical fertilizer) and maintaining soil fertility in the region. The basic idea under such efforts is to create a system of regional-based material cycles, with cooperation between livestock farmers and cash grain farmers within each region or locality. A number of research projects have also been started to examine the sustainability of alternative land use systems by researchers of the government agricultural research institutes and universities.

Southeast Asia

Southeast Asia (Thailand, Philippines, Indonesia, Malaysia, Vietnam and Cambodia) is one of the world's most densely populated regions. Its agriculture includes both highly resilient production systems that have responded positively to prolonged intensification, and extractive systems that have generally experienced drastically declining productivity over large areas. The rapidly increasing human populations, and the exploding economic expectations of the nations in the region, forebode enormously increased future pressures on the productive capacity of all land use systems.

A research base appropriately targeting the most critical constraints to the sustainability of current land use systems will be a critical component of future national well-being. A large proportion of all agricultural research is engaged in the development of specific technologies that will contribute to increasing the productivity of production systems. However, it has become increasingly apparent that research institutions may also need to focus some work explicitly on the formulation and evaluation of sustainable land use systems (Hart and Sands, 1991; Conway, 1986).

Agricultural policy in the Southeast Asian nations is dominated by concern for an adequate rice supply. Most countries are nearly self-sufficient in rice production, or are exporting surpluses. Government price supports, where they are practiced, are generally low, favouring inexpensive rice supplies for urban populations. Although the majority of the working population in all countries is agricultural, the domestic terms of trade generally favour the development of the industrial sector of the economy, indirectly taxing agriculture. Farm size is typically quite small-scale (0.5 to 3.0 ha), although a significant corporate plantation crop sector is present in some countries. Today, the market economy has penetrated to the more remote areas in nearly all countries.

Much work is in progress to elevate rainfed rice soil organic matter and build up crop nitrogen supply with precise green manures. The Department of Land Development, the Rice Research Institute, the Farming Systems Research Institute, and Khon Kaen University are particularly active in developing appropriate systems in Northeast Thailand, an extensive contiguous rainfed rice area (3.5 m ha) with soils having the lowest fertility in Southeast Asia. National programmes in the other countries are also involved, many through the ARFSN and INSURF networks.

A number of research institutions in the region are engaged in serious research to develop sustainable small-holder food production systems. The work has two major facets: Research on technology and its fit within the constraints of current farming systems, and research to overcome the social and institutional constraints to implementation of appropriate forms of land tenure and group organization.

A major sustainability issue confronting forest plantations is the reduction of species and biological diversity. In the past, experience with monoculture forest plantations has proven costly. An example is the devastating effect of psyllid infestation on *Leucaena* spp. Another is the concomitant effects of fauna reduction and soil nutrient imbalances.

The resource base in most countries in the region is fast degrading. Research priorities should be concentrated on those systems with the lowest prospect for positive trends in sustainability attributes. In this context we consider there to be two priority ecosystems: The coastal lowlands and the uplands. The very low level of research on sustainability problems of the coastal lowlands is of great concern. The richness of the fisheries resources of the region is largely dependent on the rich and intact coral reefs and mangrove areas. The present trend, however, indicates a general decline in fisheries. This is mainly due to two factors: (a) destruction of mangrove and coral habitats due to indiscriminate mangrove cutting and (b) blast fishing and other improper fishing methods.

A sustainability problem that is of utmost importance for coastal lowlands are the upland-lowland-coastal interactions. Very few studies are being conducted in this area, where cost accounting both in terms of ecological and economic costs must consider the three-way interactions. Accelerated research should be stimulated as a major international priority.

The upland ecosystem of Southeast Asia presents the most difficult and alarming challenge to sustainability. The subsistence crisis of the uplands is interconnected with national political, economic and ecological stability. The overall pace of resource degradation in the uplands is drastic. A number of institutions are involved in limited components, but the interactions are so great among subsystems in the uplands that satisfactory solutions are not easily forthcoming.

Garrity and Sajuse (1992) suggest that the subsistence crisis of the uplands in farming and forestry in Southeast Asia need further institutional innovations, addressing the systems research and development issues on an international and interdisciplinary basis. The upland ecological conditions of the Southeast Asian countries are similar, and their resource management issues recur across the region, transcending national boundaries.

STATUS OF SUSTAINABLE AGRICULTURAL SYSTEMS IN AFRICA AND MIDDLE EAST

Eastern Africa

The region of East Africa constitutes a geographical area that includes Kenya, Uganda, Tanzania, Rwanda and Burundi. The climate of East Africa is tropical. Soil erosion is very serious all over East Africa. Gully erosion is the type that many countries like Rwanda, Burundi and Kenya are actively fighting. There are government laws that prohibit cultivating steep hill sides and river banks. Soil erosion control measures including terracing by farmers are widely used in Kenya, western Uganda, Rwanda and Burundi. In Kenya, gabions are built by community labour to control gully erosion. Soil and water conservation policies and by laws exist in all East African countries. However, these are generally poorly implemented.

Because animal manures contain virtually all plant nutrients, most of them in adequate levels, crop responses to them are excellent. Goat manure exhibited significant positive residual effect on crop yields. It is therefore, strongly recommended that more research be conducted on crop/livestock/soil interactions in more East African countries, and the beneficial economic impacts to the farmers should be carefully quantified.

Soil erosion control measures must be strongly emphasized. There are by laws which when enforced should go a long way in minimizing soil erosion. Rwanda, Burundi, western Uganda and eastern Kenya are good examples where farmers practise terracing whenever they cultivate hillsides. High soil acidity must be reduced through appropriate soil management. Negative effects of soil acidity seem to be significantly reduced when organic fertilizers like animal and green manures are used. Liming is not practical for small scale, poor resource farmers.

Better harvesting of food and cash crops must be devised that return as much as possible crop biomass into the soil to improve soil fertility. Mining of soils by removing large quantities of plant nutrients are removed with the biomass. A practice like burning of sugar-cane in estate plantations before harvesting must be stopped because ashes that fly away carry with them a lot of plant nutrients. Better moisture utilization by crops should be emphasized. This may be done through better application of soil and water conservation, or by using crop cultivars that have been developed for the specific environment. For example drought resistant cultivars for arid and semi-arid areas. Finally there must be proper integration of crops, trees and livestock in a manner that they positively interact with the soils and hence increase productivity of crops and livestock (Moses Onim, 1992). In eastern Africa, the most urgent environmental degradation issue is the increased movement of people into arid and semi-arid areas.

Western Africa

Western Africa comprises Burkina Faso, Chad, Gambia, Mali, Niger and Senegal. In this region most of land use systems are resulting in a continuous degradation of the natural resource base. A first step to be taken in order to reverse the actual degradation trend is to increase at all levels the awareness of people and to motivate them for a greater commitment to sustainable land use. To be efficient the traditional top-down approach must give way to an approach where researchers, extensionists and farmers will actively interact. Research also must be multidisciplinary and conducted within a team. Water and soil fertility maintenance constitute the main existing constraints for all land use systems in WAS region. According to Badji and Seve (1992), the opportunities for sustainable land use in this region are:

1. Improved use of soil and water resources: soil and water management, improving soil fertility, small scale irrigation;
2. Improved cropping practices: intercropping, relay cropping, crop rotation, home gardens, agroforestry;
3. Genetic improvement: crop breeding, animal breeding;
4. Improved use of mixed crop/livestock systems using small ruminants, animal traction, aquaculture; and
5. Improved systems to reduce yield losses: integrated pest management post-harvest technologies, improving animal health.

The Middle East and North Africa

The dryland areas of the Middle East and North Africa are a major source of food and fibre for millions of people. However, current yields are very low compared to yields of the same crops in developed countries. The main reasons for this are: (i) traditional production practices are used by most farmers and the rate of adoption of improved technologies is very low; (ii) improved soil and water conservation methods are not being implemented; (iii) farmers do not accept new technologies easily, mainly because of economic reasons; (iv) soil erosion which leads to loss of productivity; (v) low soil fertility and inadequate use of fertilizers; and (vi) low and erratic rainfall.

Strategies to increase food production and its sustainability in this region, need to include options consistent with the different land and capital resource levels of its farmers. Concern for improving the long-term productive capacity of soils, the sustainability of crop and livestock production, has generated increased interest at the national and regional levels (Francis *et al.*, 1986). More recently, however, research is being focused on resilient species, minimum effective input rates, low risk of failure, adequate minimum production in poor years, and diverse cropping patterns (Bay-Peterson, 1986; UNDP, 1989; Jaradat, 1989).

Improved agronomic practices are highly needed to increase and sustain crop production. The most important are agronomic practices focusing on soil and water conservation, increased biomass production and cropping systems to maintain soil cover and organic matter levels. Consequently, and in most countries, forage/livestock systems should be given a high priority.

Small farmers in the region are resource-poor and will find it difficult and risky to adopt a package of practices even if its potential benefits are demonstrated at a research level. Technological recommendations must allow these farmers flexibility in choosing production practices that best fit their particular environment (Jaradat, 1992).

STATUS OF SUSTAINABLE AGRICULTURAL SYSTEMS IN SOUTH PACIFIC REGION

Australia

Australia is the world's driest continent. About one-third of Australia is classified as desert and this region is bordered by a large area of semi-desert. Apart from the tropics, with its monsoonal influence, and small areas at the south-eastern and south-western extremities, no part of Australia is entirely drought-free. About 30% of the continent receives less than 20 cm of rain a year and a further 20% less than 38 cm. However, Australia's northern, eastern and south-eastern areas, with climates ranging from tropical to temperate and Mediterranean types- usually are well watered, with regular rainfall. These areas have become the main agricultural regions.

Australian economy has grown over the 200 years since settlement, essentially based upon primary production first wool and timber, then wheat, sugar, meat and other livestock products. More recently minerals such as iron ore, coal and copper have become primary export earners. Because of the ratio of population to resources, almost 80% of primary production is exported, with mineral resources contributing 58% and rural produce 42% of export revenue.

The early awareness of the potential benefit of pasture legumes contributed to the rapid development of pasture improvement technology. The ley farming system, rotation of crops with legume-based pastures, spread rapidly in the wheat-sheep zone of southern Australia following World War II. In contrast, the northern pastoral zone relied on the native vegetation as the sole source of fodder for grazing cattle under an extensive system of land use. The beef and sheep zone in the plateaus and eastern slopes of the Great Dividing Range in south-eastern Australia and the south-western section of Western Australia, have improved pastures and produce wool, prime lambs or high quality beef and veal. The other major land use zones are the dairying zone, sugar plantations and the irrigated farms for crop and pastures.

By a combination of resourcefulness and yield-improving technologies, Australian farmers have continued to achieve growth in productivity of the agricultural sector, in the face of economic pressures, particularly in the past few decades. These productivity increases helped in compensating for the steadily declining terms of trade. The short term benefits of such growth have been at the expense of some long term costs. The decline in quality of the environment, and in particular, the question of land degradation through non-sustainable farming practices have become issues of national importance.

Burch *et al.* (1987) provided a comprehensive description of the causes of land degradation in Australia and related it to the following:

- intensification of land use
- the degree of land clearing
- tillage practices
- soil acidification of ley pastures
- irrigation pollution
- chemical pollution
- fire management; and
- stock management.

A review of past and present research strategies to address the question of sustainable land use shows that conservation of the soil and enhancement of its productivity have been, and still are the major themes in research efforts. Sustainable technologies to substitute expensive and damaging inputs are being developed to be incorporated in the cultivation and crop management practices of farmers. Despite the increased awareness and concern for the condition of the land, the adoption of these technologies by farmers has been slow.

The Australian experiences in researching sustainable land use systems have much to offer. Significant developments have been made in understanding the unique features of the environment and devising technology appropriate to prevent further land degradation. According to Sriskandarajah (1992) community participation in caring for the land and collective research strategies to alter the ways of contemporary agriculture will achieve more sustainable futures for the people of the region.

STATUS OF SUSTAINABLE AGRICULTURAL SYSTEMS IN NORTH AND SOUTH AMERICA

Canada

Less than a quarter of Canada can support commercial timber production, but much less than this is capable of economic access and harvest. Only 11 per cent of the total land area is suitable for agriculture, and only 5% of Canada's area is capable of crop production. Limitations for crops are mainly due to climate and topography, both of which change considerably from north to south.

Canada's agriculture depends heavily on export markets, especially for grains, oilseeds, meats, live animals, dairy products, and potatoes. According to Manning (1986), the following land use system problems need to be addressed:

- Canada has long had the luxury of ignoring the planning and the management or its land resource.
- Canada has no vast agricultural reserves remaining.
- Canada is intensifying its use of the land remaining in agriculture. Any increases in production depend upon continuing intensification of its best land.
- While intensive agriculture is more productive, it also requires increased investment in management to permit sustained yield. More intensive agriculture is often more fragile.
- The degradation of agricultural land is widespread and includes:
    - salinization
    - loss of organic matter
    - compaction
    - acidification (exacerbated by acid rain)
    - wind and water erosion
- The urbanization of the very best land is continuing.
- Canada has no national plan. The policy response is growing, but while there are broad policy actions taking place in most jurisdictions, what we have is policy, not strategic plans with respect to the allocation of resources and with respect to the strategies for their sustained management.

United States of America

As with Canada, urbanization of some of the best agricultural land has become a problem in some regions, particularly on the east and west coasts, and other pockets of dense population pressure. In addition to loosing land, urban pollutants are probably affecting crop yield, since higher levels of ozone and acid rain are detected in some areas. Heavy dependence on fossil fuel, and the inefficiency of energy use in agriculture, including dependence on fossil fuel derived inputs such as nitrogenous fertilizer and pesticides have been mainly responsible for these problems. In addition, leaching of pesticides and nitrate into private and public drinking water supplies has become a national problem.

Soil and nutrient run-off into surface water has become a problem in many regions, with efforts for clean-up focused on particular watersheds, such as the Chesapeake Bay in the east, and the Great Lakes in the north. Abandonment of some land due to salinization (west coast) and oxidation of organic matter (muck soils in Florida and the Northeast), are also problems, as is the declining water table in aquifers used to irrigate the western Great Plains. Modern agriculture has encouraged high capital investment by U.S. farmers, and many feel locked in to their current production practices because of the necessity of yearly cash flow and debt servicing. Many also farm rented land, and are reluctant to make permanent improvements on land they do not own (*i.e.* windbreaks, terraces), and also tend to farm according to the desires of their landlords, and are not entirely free to try new practices or experiment on this land.

More research falls into the categories of pest management and cropping systems research, with the least in farmer integration, economic and policy analysis, and livestock (Janke and Blew, 1992).

Central America

Concern for sustainability results from the obvious degeneration of the natural resource base which certain of the land use systems entail. While population was low, the damage to the resource base was minimal. It has been estimated that 2/3 of the cutting of the original primary forest in Central America the cutting of the original primary forest in Central America occurred between 1950 and 1985 (Leonard, 1985). The area of natural forest in central America was 238,640 km2 in 1970 and 202,420 km2 in 1980. While there are still considerable areas of forest in central America, most of these are in unpopulated areas where they do little to reduce soil degradation or serve as a source of fuelwood in heavily populated areas. The greatest loss in forests occurred in Guatemala, Honduras, Nicaragua, and Costa Rica while areas in pasture increased more than areas in crops.

Some of the land use systems presently used in Central America may be defended as sustainable (Kass *et al.*, 1992). Others obviously are not. Much depends on the nature of the resource base. Shade grown coffee and cocoa can be maintained for many years with no obvious degeneration of the resource base. At low population densities, shifting cultivation is a sustainable land use system. On some of the better soils of the central highlands of Guatemals or the Pacific coast of Central America, even annual crop production as presently practiced, has not resulted in any obvious degeneration of the soil resource. It may be argued that considerable latitude is permissible in these areas because not only is the soil resource of high quality, it is periodically renewed by volcanic activity (Kass, 1981, 1983). One way or another, there are areas around Guatemala City which have probably been in continuous maize cultivation for more than 2000 years. Historically, Central America has much to offer in considerations of sustainability. The lowland Maya are usually considered a prime example of a civilization that declined because of a degeneration of its resource base. On the other hand, the highland Maya, with admittedly a perhaps superior resource base and less problematic climate, seem to have achieved a sustainability of sorts.

Unfortunately, in areas less favoured, degeneration of the resource base is obvious and productivity has not been maintained. Because of the existence of more economically profitable uses of the flatter areas with deep soils, food production has moved increasingly to steep and shallow soils which may not only be chemically depleted but may no longer exist after two or three cropping cycles. Making crop production on such sites sustainable would certainly be a valuable contribution. Unfortunately, the return obtained from cereal crops would not finance the inputs necessary to make such sites sustainable. The introduction of higher value crops are more likely to furnish such inputs. Sustainable production systems for these high value crops urgently need to be developed.

Agroforestry has been frequently proposed as a means of making annual crop and pasture systems sustainable (Sanchez, 1987). While the introduction of leguminous trees into crop and pasture systems could bring the same benefits as they have brought to coffee and cocoa plantations, export of nutrients from crop and pasture systems is generally higher than in beverage crops (Sanchez, 1976).

In recent years, increasing population densities, desire for increased profitability of land uses, and the need for export earnings have resulted in modifications of these originally sustainable systems with far-reaching environmental consequences (FAO, 1975; Leonard, 1985). Multiple cropping systems which occupied flat areas and used high levels of organic residues were displaced to less favoured sites as more favoured sites were used for more profitable activities: lowland monoculture of crops mostly used for export and extensive cattle grazing (Hawkins, 1985). The thin soils on hill slopes rapidly degenerated under practices designed for more highly favoured soils. Shade was removed from coffee to obtain greater productivity and, to a lesser degree, to spray for coffee rust, with less protection of the soil resource (Budowski *et al*, 1984) to maintain productivity and to exploit the forest resource, more than half of the existing forest has been destroyed since 1950 (Leonard, 1985).

STATUS OF SUSTAINABLE AGRICULTURAL SYSTEMS IN EUROPE

Eastern Europe

All the countries of the region with the exception of Poland have large area collective and state farms. In Poland large state and cooperative farms account for only slightly less than 20 per cent of the agricultural land. The remaining 80 per cent of the farmland is managed by small and medium sized traditional private farmers, their area varying from 1 to about 50 hectares.

In almost all countries of the region a greater or smaller deficit (shortage) of farm products is found. The basic target of the central administration and of the political factors in these countries was the continuous intensification of agricultural production, striving to create bigger and bigger area units, which were easier to manage centrally and excessive concentration of livestock production (the so called industrial feedlots for meat production). These big livestock farms result in great losses of organic fertilizer and cause severe pollution of the natural environment.

According to Drupea (1992), all of the enumerated trends have disrupted the ecology of agricultural lands, resulted in serious pollution of soils and ground and surface waters, causing soil degradation processes, thus in general were and remain still in conflict with the principles of sustainable agriculture.

Western Europe

Western Europe is a conglomerate of countries with differences in climate, history and agricultural traditions. In Northern Europe, milk production, pastures, wheat, barley and sugar-beets are the characteristic elements while in alpine valleys, milk and cheese production, woodlands and pastures dominate. In Mediterranean Europe, citrus and olive orchards, vineyards, vegetables, cereals and pasture, especially on the slopes are dominant. In the lowlands valleys (for example the Po valley) some cash crops, such as corn, wheat, soybean and sugarbeets are grown. Cattle, poultry and pigs are raised without sufficient land for sludge disposal.

Sustainability of the different parts of this scenery pose serious questions. Energy, technology, pesticides and chemical fertilizers are being increasingly overused in most parts of the area. Dependence on external input, soil erosion, pesticide resistance, ground water pollution, lake, river and lagoon eutrophication and food contamination are the main claims for changes, as well as decreasing incomes for small farmers and, in certain cases, over-production (milk, certain cereals).

It has recently been observed that the IPM (Integrated Pest Management) European projects have had limited success among farmers (Vereijken, 1989; El Titi and Landes, 1990) due to the partial objective of the classical IPM projects to control pests on a farm in which most parameters are not ecologically designed or performed (Altieri, 1987; Edwards, 1990).

The options for sustainable agricultural systems are to handle a farming system in order to minimize pests as well as costly inputs of chemicals and energy and to optimize rotations and organic matter in the topsoil (El Titi, 1990; Vereijken, 1990). Seen from this viewpoint, for instance, some problems of soil pests in sugarbeet (in Germany) can be indirectly controlled with reduced tillage, nitrogen localization and a post-emergence weed control strategy in spite of chemical insecticides applied at seedling time or the high rate of chemical herbicides (El Titi, 1990).

Sustainability must be developed as an integrated strategy in connecting primary production with the new needs of the post–industrialized society. Emphasis need to on maximum quality of the productions instead of the maximum quantity, and maximum environmental-compatible quality which implies higher biodiversity (Paoletti, 1992).

CHAPTER 3

Management of Agricultural Pollution: Soil Pollution

Man is dependent on soils and to a great extent soils are dependent upon man taking into consideration the use he makes of them. Soils are natural bodies which support plant growth. Any action on the part of man or soil which contaminates the soil plant ecosystem can lead to soil pollution which may have variable effect on soil health and plant growth.

Pollution contaminates things and makes them unclean or impure. Soil and water are polluted in many ways. Physical degradation results when eroded soil becomes a water pollutant or buries a fertile soil beneath unproductive sediment. Soil may be covered by discarded trash or ruined chemically by contamination with heavy metals or other toxic materials. Poisons sometimes escape from soil or trash into streams and lakes where they kill fish and perhaps birds and animals that live on the fish. The soil may be biologically degraded either by toxic materials or by poor management that exhausts its fertility, depletes its organic matter, and changes its soil structure. The pollution problem can no longer be ignored because there is no fresh desirable land to replace that which is worn out. Earth's resources are now recognized as being a closed system of limited size that is much easier to damage than to improve. Soil and water pollution squander the resources that support life.

People have caused pollution for as long as their history can be traced. In fact, the careless discards of bygone civilizations have yielded much of the available information about ancient people. There was relatively little concern about pollution until recent times, however, because the population was only a fraction of its present size, there was plenty of room for both the people and the pollution they caused. The earth is now so densely populated that space restrictions limit the standard of living in many areas. Reducing productivity by soil pollution further decreases the standard of living. Water pollution has many unpleasant effects including aesthetic, health, and productivity problems.

The damage caused by pollution is difficult to evaluate accurately but some estimates are available. Shaw, Heggestad, and Heck (1971) state that pollution damage in the United States adds up to many billions of dollars each year. They indicate that air pollution alone causes more than half a billion dollars of damage to agricultural crops annually. That loss might double by the year 2000 because increased crop production will be required. Pollution itself will increase unless billions of dollars are spent to control it.

Another reason for increased concern about pollution is the variety of new types of pollutants. Items made of wood and other plant materials and those composed of animal products cause little pollution. The smelting of ores for metal products has caused some pollution for a long time and a growing amount since the industrial revolution. The wide variety of useful products coming from chemical industries in recent decades has been accompanied by an equal variety of pollutants. The more exotic materials are more likely to become pollutants rather than decompose into harmless chemicals. Fortunately the trend towards more serious chemical pollution is now being slowed or reversed by considering decomposition rates when new products are developed.

Nuclear wastes are the newest major class of pollutants and the cause of much concern. Part of the problem is the very long lifespan of many radioactive wastes and the extreme toxicity of some materials, especially plutonium. Safe disposal of such materials is a difficult problem that has not yet been solved on a long-term basis. Also, what should be done with the power plants when they are worn out or become obsolete?

The increasing concern about pollution has been accompanied by more sensitive means of detecting pollutants and measuring their concentrations. Parts per trillion are measured as routinely now as parts per billion were measured a generation ago and as parts per million were before that. This increased sensitivity identifies pollutants that would have escaped notice before and raises the question of how much hazard such tiny amounts can cause. Tolerable levels may need to be defined in some way other than the minimum detectable concentrations.

Here, only the important aspects of soil pollution have been briefly summarised. These aspects are also discussed in detail in the following chapters.

SOIL POLLUTION

Some pollutants are natural materials that either get in the wrong place or become too concentrated in some places. Soil can become a pollutant by getting into water or air. Soil is also a part of the environment that receives pollutants. Soil can depollute some materials by causing them to decompose into harmless end products such as carbon dioxide and water. Useful plant nutrients are often released by decomposition of pollutants. But there are some materials that are difficult or impossible to decompose. Accumulations of such materials can pollute the soil. Soil pollution therefore includes three different aspects; soil as a pollutant, soil as a depollutant, and polluted soil.

Soil pollution often relates to diffused sources. Leaching, volatilization, sediment, and dissolved material in runoff are all nonpoint sources of pollution. Extensive soil conservation and other good management practices are the best means of limiting such diffused pollution. Spreading waste material across a wide area allows the soil to depollute the waste even though it would pollute a smaller volume of soil.

Soil as a Pollutant

Soil is the source of dust in the air and sediment in water. On the basis of tonnage, soil is by far the largest pollutant. The 4 billion tons of soil erosion occurring each year in the United States of America is hundreds of times greater than the tonnage of sewage entering that country's streams. The largest fraction of the soil is inert material, but even inert material can fill a reservoir. Eroded soil carries with it bacteria and viruses, but disease problems from this source are unlikely.

The Indira Gandhi Canal in India runs through Sriganganagar, Bikaner and Jaisalmer districts of Rajasthan. The canal system is facing an alarming danger from blowing and drifting sand which causes sand deposition inside the canal and adversely affecting the flow of water. The main canal has been constructed by cutting massive sand dunes and there is practically no vegetative cover to check the drifting sand. The height of the sand dunes varies from 5 m to 65 m above the level of the canal. The recurring problem of wind deposited sand is of a very serious magnitude as nearly half a million cubic metres of blown sand had to be removed from the main canal from a 61.6 km stretch of the canal. The sand drift has been estimated as high as 1400 to 5500 kg/ha during summer months. Such a sand erosion poses a threat not only to the canal system but also to the agricultural development envisaged in the canal command area. The wind velocity varies from 5 km per hour to 27 km per hour in different months and during the dust storms, the wind velocity may exceed even 100 km per hour. Generally, the minimum threshold velocity ranges from 5 to 15 km per hour for checking/containing the soil erosion. The alignment of the main canal is from North-East to South-West and the wind direction is just opposite to this, and hence, maximum deposition takes place on the left bank and left side of the canal. In the wind erosion process, the sand moves by suspension, saltation and surface creep. The phenomenon of sand movement is intimately connected with the velocity of the surface wind, the force exerted by it on the sand grains and also the resistance offered by the grains to the passage of the wind. The moving particles may be deposited from the wind by way of accretion, encroachment and sedimentation. Samples collected from sand dunes adjacent to left bank from different locations were subjected to mechanical analysis. The grain size analysis of samples indicate that these sand dune samples as well as samples of accumulated sand in the canal possess predominantly fine sand particles as that of embankment materials. Moreover, the grain size distribution of the dune and canal deposit at the same location shows more or less the same characteristics which indicate that the main source of the sand deposition in the canal is predominantly the sand dune.

Plant Nutrients

The plant nutrient content of soil is usually its most important characteristic as a pollutant. The growth of algae and other plant life in bodies of water is usually limited by the supply of phosphorus or, less often, of nitrogen. Soil contains both of these nutrients and can supply them to streams, ponds, and lakes through leaching and by runoff water carrying both dissolved material and sediment. The result is often eutrophication. Sediment usually carries considerably more nutrients than the amounts dissolved in water. An especially large percentage of the phosphorus moving from sod to water is carried by sediment because most phosphorus compounds have low solubilities.

Phosphorus is the nutrient most likely to be deficient enough to limit algal growth in water. Adding a phosphorus supply is therefore the most common way of stimulating unwanted growth. Phosphorus is also the nutrient most amenable to control when efforts are made to improve the quality of water. Erosion control practices such as contouring, terracing, and border strip-cropping can keep much sediment out of streams. The concentrations of phosphorus and other nutrients associated with the sediment are lowered accordingly. At least two other sources of phosphorus also need to be controlled to keep water clean. One is runoff from feedlots, which are too often placed next to streams. The other is sewage that has had only primary and secondary treatment. The effluent from secondary treatment contains large amounts of phosphorus coming from detergents and various organic wastes. Tillage and fertilization also have much to do with the amount of phosphorus moving from land to water. Reduced tillage reduces sediment, but the reduction may be partly offset by increased soluble phosphorus in the runoff resulting from surface applied fertilizer. Proper rates of fertilization are important in any cropping system because too much fertilizer results in excessive nutrient loss to streams. The right amount of fertilizer, however, can improve growth and reduce erosion enough to reduce the amount of sediment and nutrients carried to streams. Proper fertilization is therefore a good way to reduce pollution.

Soil loss from forest land is normally less than that from cropland.

The greatest hazard comes when the forest is harvested. Even then, the nutrient loss need not be large, nor should forest fertilization cause much increase in nutrients carried by runoff. Forest fertilization rates are seldom very high and the runoff water passes through too much litter material to carry large concentrations of nutrients.

Soil as Air Pollutant

Wind erosion contaminates the air with large amounts of dust that maybe carried hundred of kilometres from the source. The dust may be a nuisance or worse wherever it settles. Respiratory problems are aggravated by dust, clothing gets dirty, machines wear out and experiments can go wrong as a result of dust contamination. The smallest dust particles accumulate in the upper atmosphere, sometimes in large enough concentrations to intercept significant amounts of sunlight. The worst episodes of wind erosion occur when dry climatic cycles strike cultivated land in arid, semiarid areas. The desolate conditions thus produced have aptly been called dust bowls. Smaller amounts of dust pollution, can occur anywhere.

Soil also contributes into the atmosphere. Nitrogen escapes by ammonia volatilization and by denitrification. Decomposition of organic materials in soil can release sulphur dioxide and other sulphur compounds. Most of the gases emanating from the soil eventually return to the soil as the atmosphere is cleansed by precipitation. Dust particles also return to earth, often as the nuclei of raindrops.

Soil as Depollutant

Many people assume that anything deposited in the soil will decompose. This assumption is not quite true. Some metals, plastics, and other materials will remain unchanged because they resist both chemical and biological decomposition. Most of these are synthetic materials; for example, stainless steel is specifically designed to resist oxidation. Some plastics are among the materials that cannot be decomposed by soil microbes because they differ too much from natural organic materials. But soil is still the closest thing to a universal depollutant. Most things will rot, rust or otherwise decompose in soil if given enough time.

Using soil as a depollutant is an extension of natural recycling that makes beneficial use of most organic wastes. Materials that would be pollutants anywhere else become energy sources for soil microbes and nutrient sources for plant growth when applied to the soil.

Waste Disposal

The fertilizer value of manure is secondary to the problem of waste disposal. Feedlot concentrations of tens of thousands of cattle and chicken concentrations of hundreds of thousands of birds in single operations produce tens of hundreds of tons of manure per day. About 100 million metric tons of animal waste (dry weight) are produced in the United States of America each year. These wastes contain nearly 4 million tons of nitrogen and nearly 1 million tons of phosphorus (Van Dyne and Gilbertson, 1978).

Feedlot operators seldom have enough land available to spread their manure at the most appropriate fertilization rates of perhaps 10 to 20 tons/ha. They may want to use rates 10 times that high so only one-tenth as much land is required. The manure concentration becomes high enough to be a source of pollution both in the pens and on the land used for disposal.

Manure emits a variety of nitrogen and sulphur compounds that pollute the atmosphere. Some of these compounds have strong offensive odours. The best way to control these odours is to spread the manure at reasonable rates and incorporate it promptly into the soil.

Manure on the soil surface is also a source of water pollution. It yields soluble plant nutrients to runoff in addition to the nutrients carried in solid particles. Incorporation into the soil helps solve the water pollution problem as well as the air-pollution problem. The possibility of leaching remains, but leaching losses are usually small compared to runoff losses.

Manure can be applied to soil in a variety of ways. Spreaders are available to distribute the solid material either while it is fresh or after it has been composted. Manure can also be fermented in a lagoon and then spread with a sprayer truck or an irrigation system. Any of these systems converts a potential pollutant into a valuable source of plant nutrients.

Larson *et al.* (1975) estimated that sewage sludge contains nutrients equal to about 5 per cent of the N, 12 per cent of the P, and 1 per cent of the K applied in commercial fertilizers in the United States of America. Significant quantities of micro-nutrients are also present in the sludge, some of them in chelated forms that help to protect their availability. Some researchers have obtained higher crop yields using high rate of sludge application than they could with fertilizer and lime, probably because the sludge improved the physical properties of the soil as well as the plant nutrient supply (Willett *et al.*, 1986). Additional nutrients that could be turned into fertilizer are contained in sewage effluent, garbage, and other organic wastes.

Sludge should usually have some pretreatment to avoid toxicity and pathogenic problems when it is applied to soil. For example, the city of Philadelphia (USA) has developed a five-step process for treating its sludge (Miller, 1986): (1) mechanical dewatering to increase its solid content from 4 or 5 per cent up to 20-25 per cent. (2) blending with 2.5 parts woodchips per part of sludge by volume to increase porosity, (3) aerobic digestion (composting) to kill pathogens and make a more mellow material, (4) curing for about 30 days to produce a drier product, and (5) screening to prepare the material for marketing and to recover the woodchips for reuse.

Favourable results from sludge fertilization have also been reported for various types of cropland (usually on forages so there is no direct human consumption of the crops produced) and on forest plantations (Brockway, 1983; Lutrick *et al.*, 1986). Very high rates of sludge (110 to 330 tons/ha) have sometimes been applied to areas of mine spoils (Elliot, 1986) with very favourable results. Such rates are high enough to provide a significant organic component and to contribute to the cation exchange capacity of the mine spoil materials.

Not all soils are equally suited for waste disposal. Suitable soils should be reasonably permeable and well aerated so that oxidation can take place. The soil should not be extremely fine or extremely coarse textured. A coarse-textured soil may pass the wastes through too rapidly for the soluble materials to be decomposed. It may be impossible to maintain adequate drainage in a fine-textured soil even if it is satisfactory at first; it is possible for the suspended solids and the biological exudates produced when sludge is added to the soil.

Sewage effluent is less hazardous than sludge for application on soil because the pollutants are much more dilute in the effluent. In fact, the effluent may be considered as an important water resource in a dry climate. After 18 to 25 years of sprinkler or flood irrigation, some soils in Israel studied by Burns and Rawitz (1981) are showing decreased infiltration rates. They suggest that drip irrigation may be the best method for using effluent because it reduces the exposure of people and aboveground plant parts to the effluent.

A forage crop should be grown and harvested where land is used for waste disposal. The crop aids in erosion control, in maintaining soil permeability, and in the removal of soluble materials that would otherwise build up in the soil and groundwater. Harvesting the forage maximizes the amount of nutrients and the ions removed from the area.

Wastes are sometimes applied at rates of 110 to 220 tons/ha and may carry as much as 1100 kg/ha each of N, K, and Ca plus proportional amounts of other nutrients. Even forage crops are not likely to remove such large quantities of nutrients. Fortunately, several soil processes provide other means of nutrient removal. The loss of nitrogen by denitrification can be encouraged by having at least two areas to alternate where an irrigation system is used. Nitrification takes place during dry periods and denitrification during wet periods. About 5 days are required to allow enough oxygen to enter the soil for nitrification of the ammonia contained in sewage effluent applied during the preceding 6 days.

High rates of waste application increase the likelihood of polluting either the groundwater or the soil. Even with alternating wet and dry periods, it is possible to leach nitrogen from the soil. The ammonium form can be leached if the wet periods are too long. The nitrate form may be leached if the wet periods begin too suddenly for denitrification to lower the nitrate concentration. Nitrate contamination of groundwater is one of the chief concerns in part because of the possibility of causing methemoglobinemia (blue baby) toxicity. Human infants are susceptible to methemoglobinemia because bacteria in their systems can reduce nitrates to the much more toxic nitrites. Humans more than 6 months old have enough acid in their digestive tract to kill these bacteria (Hergert, 1986). For drinking water, less than 10 ppm NO_3^--N is safe.

Heavy applications of wastes increase the salinity of the groundwater under any conditions. Salts such as sodium chloride and calcium chloride are too soluble to be prevented from leaching. Usually the increased salinity is not a problem unless the water also receives large salt inputs from other sources.

Element Toxicity

Heavy metals and other potentially toxic elements are the most serious soil pollutants in sewage. Bradford et al. (1975) found toxic levels of boron and excesses of several heavy metals in some plants irrigated with sewage materials. The heavy metals are a matter of concern because they accumulate on cation-exchange sites and are very difficult to remove from the soil. Harvesting forage crops removes small quantities of the heavy metals as long as the concentrations are below toxic levels. Still smaller amounts of heavy metals are emitted to the atmosphere in particulate matter while plants are growing.

Sewage sludge contains higher concentrations of heavy metals than most other wastes contain. The concentrations vary widely from one community to another and from time to time, but they are consistently high enough and that heavy metal concentrations should be monitored where repeated heavy applications of sludge occur. Larson *et al.* (1975) recommend that the zinc equivalent not be raised above 5 to 10 per cent of the soil's cation exchange capacity.

Some soils have been contaminated with heavy metals from manufacturing wastes that reached them either through drainage water from a factory or from materials discarded in some kind of a trash dump. Less commonly, but equally serious where it occurs, aerial emissions can contaminate soils. Drozd, (1984) report on contamination from a copper metallurgy factory that emitted large amounts of copper, zinc, and sulphur into the atmosphere. The copper concentration in the soil was more than 100 times normal for a distance of about 1 km downwind, and more than 10 times normal for 2 km. Plant cover was killed and the soil left bare in the worst part of the affected zone.

Polluted Soil

Soil can be overloaded in its role as a depollutant and become polluted itself. The productive potential of the soil may be reduced or eliminated, or plants growing on the soil may absorb toxic materials that cause problems at some point in the food chain. The damage may be either short lived or long lived. Many pollutants are so costly to remove that it becomes uneconomical to reclaim the land.

Excess amounts of some pesticides may kill or injure plants, insects, and animals that were not intended to be targets. The overload may make the pesticide more persistent than normal by killing the microbes that would otherwise decompose it. Sometimes the chemical applied is converted to a related form that is also detrimental.

Salinization

Excess quantities of wastes can cause a buildup of soluble salts in the soil. Part of the salts are carried in solution in the liquid portion of the wastes and others are formed as the solids decompose. Salts accumulate in the soil when the amounts added in wastes exceed the amounts removed by plant growth, leaching, and other means. Eventually a saline or a saline-sodic soil may be produced. The worst such condition occurs when leaching removes the excess salts but leaves enough sodium to form a sodic soil.

Soil pollution is widespread and comes in many forms. The solutions are seldom simple but pollution can be controlled by concerted effort. It is possible for man to use soil productively without degrading it.

Saline and sodic soils are occasionally caused by other forms of pollution. For example, in places where salt is used to melt snow and ice, there may be enough salt entering the soil along roads and sidewalks to sterilize the soil. Usually the affected area is only a narrow strip and it may be only an occasional spot along the way. Salty water dumped after freezing ice cream may also kill vegetation and damage the soil. A similar problem arises when seawater floods land because of a tidal wave.

Heavy Metals

Soils normally contain small amounts of heavy metals like lead, zinc, copper, and mercury. In fact, copper, iron manganese, molybdenum, and zinc are included in the list of essential elements. But excessive amounts of these and other heavy metals are detrimental to life and can sterilize the soil. Heavy metal pollution is serious because it can persist for many decades.

Heavy metals reach the soil in several ways. Small amounts cycling through plant growth and returning in residues are normal and cause no problem. Larger amounts of heavy metals contained, for example, in repeated large applications of sewage sludge can be too much. Some industrial wastes contain such high concentrations of heavy metals that they should neither be spread on land nor dumped in water.

Lead and other heavy metals used in fuels and lubricants find their way into the atmosphere and drift across areas near heavily traveled routes. Air, water, and soil may all suffer pollution from this source. MacLcan and Langille (1973) reported substantial increases in the lead, copper, and molybdenum contents of soils and plants along roads. Bordeaux mixture (copper sulphate and lime), various mercury compounds, and other preparations containing heavy metals were once the only kinds of fungicides and insecticides known. Applications had to be many times as heavy as modern pesticides and often had to be repeated every week or two. The soil beneath some fruit trees became heavily contaminated with heavy metals that dripped or washed from the trees after they were sprayed. Plant growth on some orchard sites is still affected even several decades after the trees were removed.

Heavy metals are difficult to remove from soils because they are strongly held on cation-exchange sites. The concentrations in solution are therefore low, and leaching is relatively ineffective for removing them from the soil. Growing plants can remove significant amounts if the metal concentration is not high enough to prevent plant growth. Sometimes the activity of a heavy metal can be reduced by adding something that reacts with it to form an insoluble compound. For example, Kunishi and Taylor (1972) were able to reduce the extractability of strontium added to soil from 80 to 100 per cent down to 2 per cent. They added diammonium phosphate, lime, and ammonium fluoride to the soil so the strontium would be converted to insoluble apatite.

Some problems caused by heavy metals are actually deficiencies of other nutrients. For example, iron and manganese are antagonistic to each other because they enter roots by the same carrier molecules. Excess manganese can therefore cause an iron deficiency. A suitable iron source applied to the plants or the soil may solve such a problem. Also, excesses of iron and several other heavy metals form insoluble phosphorus compounds and can create a phosphorus deficiency. Phosphorus fertilizer and liming may solve the problem.

CHAPTER 4

Management of Agricultural Pollution: Agrochemicals

At present, thousands of chemicals are used for agricultural purpose. Some of these chemicals are being used in large quantities to cover cropped areas and livestock. These chemicals interact with different components of the plant ecosystem and contamination may take place in respect of one component or the other. In recent years the use of agricultural chemicals especially fertilizers and pesticides has become a cause for concern as some have been marked out as pollutants having adverse effects on environment including human beings.

According to recent estimate over 80 million mt of N, P and K are being used annually in the world in the form of chemical fertilizers. Fertilizer use is following an increasing trend as they are potent source of increasing crop production and meeting the demand of the growing population. However, some people who support the organic or biological system of agriculture believe that the food and other needs of man could be properly met by minimising the use of synthetic fertilizer.

In recent past several reports suggest that the agrochemicals, especially the chemical fertilizers are resulting in environmental degradation by way of polluting plant and soil environment, water bodies, effecting the ozone layer, etc.

AGROCHEMICALS AND THEIR IMPLICATIONS ON SOIL ENVIRONMENT

A wide range of agrochemicals are currently used by farmers to sustain food production. These chemicals range from fertilizers that include N, P and K to pesticides to control both weeds and pests (Table 4.1). With the exception of macro and micronutrient cations, excessive use of both anionic fertilizers (phosphorus and nitrate) and pesticides has been shown to have adverse impacts on soil and water environment (Hester and Harrison 1996). An additional all constraint with fertilizer use has been the inadvertent addition of trace metals such as Cd and Hg that has implications on both soil and crop quality. The most alarming issue facing scientists and the community is the disproportionately high use of agrochemicals for crop production compared to establishment of research activities and regulatory measures to protect soil environment

TABLE 4.1: Commonly used agrochemicals

(A) Fertilizers

Nutrient	Fertilizers
Nitrogen	– Ammonium chloride
	– Ammonium sulphate
	– Ammonium nitrate
	– Urea
	– Calcium cyanamide
	– Sulphur coated slow release fertilizer
	– Ammonia gas liquors
	– Aqueous ammonia
	– Anhydrous ammonia
Phosphorus	– Single super phosphate
	– Triple super phosphate
	– Monoammonium phosphate
	– Diammonium phosphate
	– Basic slag
	– Rock phosphate
Potassium	– Muriate of potash
	– Sulphate of potash
	– Potassium nitrate

(B) Pesticides

Group	Type
Quaternary N pesticides	– Diquat, paraquat
Basic pesticides	– Ametryn, amitrole, atrazine, benomyl, cyanazine, metribuzin, prometryn and simazine
Carboxylic acid herbicides	– 2,4-D, dalapon, dicamba, dichlorprop, endothal, MCPA, picloram
Hydroxy acid pesticides	– Bromacil, bromoxynil, dinoseb, terbacil,
Ammonosulfonyl acid herbicides	– Asulam, chlorimuron, chlorsulfuron, metsulfuron, triasulfuron
Organic As and P acid pesticides	– DSMA, glyphosate, MSMA
Organometallic fungicides	– Mancozeb, maneb, metiram, zineb, ziram
Amide and anilide herbicides	– Alachlor, metachlor, napropamide, propachlor, porpanil
Carbomate pesticides	– Aldicarb, carbaryl, carbofuron, chloropropham
Chlorinated hydrocarbon pesticides	– Chlorothalonil, endosulfan, lindane, metoxychlor
Organochlorine pesticides	– Aldrin, chlordane, DDT, heptachlor
Organophosphate pesticides	– Chlorpyriphos, diazinon, fenamiphos, malathion, monocrotphos, parathion, phorate, phosmet, terbufos
Phenylurea pesticides	– Diuron, flumeturon, linuron
Pyrethroid pesticides	– Bifenthrin, cypermethrin, fenvalerate, permethrin
Thiocarbomate herbicidet	– EPTC, molinate, trillate

ROLE OF AGROCHEMICALS IN INTENSIVE AGRICULTURE

Global fertilizer use has shot up from 14 million to 125 million tonnes, an increase of almost 900 per cent (Saull 1990) over the last 30 to 40 years. In Australia alone, over 15 million tonnes of superphosphate has been used during the last 100 years. Calculations, based on the Cd content of rock phosphate used to manufacture fertilizer show that it is estimated that over 6000 tonnes of Cd have been added to the Australian soils through fertilizer use in the last 90 years (McLaughlin et al., 1996). During 1950- 70, growth in world fertilizer use was mainly concentrated in the industrial world. In the developing countries, significant use in chemical fertilizer did not begin until the sixties demand for food driven by population growth and expanding appetite for animal protein has led to massive applications of fertilizers in many developing countries as well. Brown (1985) reports that as of 1981, the industrial countries were consuming 72 million tonnes of chemical fertilizer, 63% of the world total. Developing countries were using 43 million tonnes, just over a third of the total, but their consumption has been growing far more rapidly suggesting that by the year 2000 usage may be rather evenly divided.

Apart from to increased fertilizer use, modern cultivation practices have also relied on increasing pesticide use. With the introduction of DDT in the early 40's, non chemical methods of pest control dwindled. Today, over 100,000 chemicals including insecticides, fungicides, herbicides, acaricides and rodenticides are in use throughout the world for pest and weed control. In the United States alone pesticide usage more than tripled in the last 30 years (Postel 1988) and on average farmers use about 2.8 kg/ha. Whilst pesticide usage in developing countries is not as high, establishment of manufacturing plants in some Asian countries has led to an astronomical increase in pesticide in the agricultural environment. Pesticide consumption in Pakistan increased from 500 a.i. ha^{-1} in 1982 to 1000 g a.i. ha^{-1} in 1990 (Hussain *et al.*, 1996). India is the largest manufacturer as well as consumer of pesticides in Southern Asia (Prasad, 1996). The production of technical grade pesticides in the country during 1991-1992 was approximately 74,000 t, whereas total consumption was 72,000 t. On average approximately 500 g ha^{-1} is used in India. Pesticide usage in India and Pakistan and other Asian countries is quite low compared to many European countries, Japan and USA. This, however, does not imply that the problem of pesticide contamination does not exist in these developing countries.

India has made an impressive progress in food production during the past four decades. The food-grain production has risen from 52 million tonnes in 1951-52 to 198 million tonnes in 1996-97. During the period 1965-66 to 1969-70 production increased by 36 million tonnes at an annual growth rate of 7.2 million tonnes per year. This increase often termed as Green Revolution, was largely due to the increase in area under cultivation, without much use of agrochemicals. On the other hand, during the period 1970-71 to 1990-91, the net cropped area virtually remained static. A spectacular rise of 67.8 million tonnes in food grain production from 108.4 million tonnes in 1970-71 to 176.2 million tonnes in 1991 has largely come from increased area under HYVs, intensive use of synthetic fertilizers and pesticides and increased area under irrigation, which enabled farmers to adopt intensive cropping with rice and wheat.

The agrochemical industry in India has come a long way. Starting with a very meagre beginning, the country now produces sophisticated pesticides, fertilizers, plant hormones and other allied chemicals to meet the food and fibre requirements of the expanding population. The rate of increase in the use of agrochemicals in developing countries is considerably higher than that of the developed countries.

The Indian agrochemical industry is developing into a major exporter. Exports almost trippled from Rs. 640 million to Rs. 2342 million during the four year period (1988-89 to 1991-92). Pesticide sales at the consumer level in the domestic market were about $ 750 million. India has 170 million hectares of arable land, including 80 million hectares of irrigated land. But only about one-third area is covered under agrochemicals.

While both fertilizer and pesticide usage has led to increased crop production and greater economic returns to farmers, their impact on soil and water environment and crop quality was not considered relevant until the last ten years. There is now increasing evidence to show that indiscriminate use of agrochemicals can have adverse effect on soil environment (Naidu *et al.*, 1996).

FATE OF AGROCHEMICALS IN THE SOIL ENVIRONMENT

Soil is a major sink of agrochemicals whether the chemicals are applied directly (e.g., soil or foliar application) or indirectly (e.g., by aerial spraying). Since most concern is associated with the presence of agrochemicals in the soil environment, it is essential that the processes affecting the fate of agrochemicals in the soil be understood. The important processes affecting the fate of inorganic (fertilizers) and organic (pesticides) ions are discussed below.

Inorganic Fertilizers

Amongst the inorganic fertilizers in use, excessive use of nitrate and phosphatic fertilizers have led to extensive contamination of surface and ground waters. The fate of N fertilizers in the soil environment are controlled by several physical, chemical and biological factors that interact with each other. The main processes affecting the fate of fertilizer in the soil are: plant and animal uptake (immobilization), adsorption and exchange in the soil (fixation), leaching or infiltration to groundwater, volatilization and gaseous losses to the atmosphere and surface loss in solid form by runoff and erosion (Ellis and Mellor, 1995). The per cent recovery of nutrient varies between the different types of fertilizers. It has been reported that 50-65, 5-15 and 75% of nitrogen, phosphorus and potassium, respectively were recovered from inorganic fertilizers (Fink, 1982).

Amongst the nutrients, nitrogen is quite often found in drainage waters, for example, the loss ranges between 15 to 150 kg/ha, whereas levels of phosphorus are generally low (< 1 kg/ha) (White, 1987). Losses upto 30-35 per cent of the applied N have been recorded for both leaching and denitrification. Leaching is most commonly found in coarse-textured and well drained soils, whereas denitrification losses are greatest in fine-textured, waterlogged and poorly aerated soils. In terms of environmental problems associated with fertilizer use, nitrate leaching has received much attention in recent years (Addiscott, 1996).

Phosphate, sulphate, nitrate and chloride are the major nutrient anions added to soils through fertilizer addition. Some of these anions are implicated in the contamination of groundwater and surface waters resulting in eutrophication of water bodies. Anions in soils are generally grouped into "reactive" and "non-reactive" anions. This is mainly based on the extent of adsorption of anions onto the soil particles. Traditionally phosphate and sulphate are included in the reactive group and nitrate and chloride in the non-reactive group. The reactivity of anions influences their persistence and mobility in soils. In general, the reactive growing interest in the potential impact of soil anions on the contamination of groundwater and surface waters.

Specific and Non-specific Adsorption of Anions

The adsorption processes of anions and cations are grouped into "specific" and "non-specific" adsorption. Non-specific anion adsorption is a process in which the negatively charged anions balance the positive charges on the soil particles through electrostatic attraction. Whereas the specific adsorption involves chemical bond formation between the anions and the ions in the soil surface. Nitrate and chloride are considered to be adsorbed through non-specific process and the adsorption of phosphate and sulphate involves both specific and non-specific adsorption processes.

Surface Charge

Surface charges in soils are grouped into "permanent" and "variable" charges. Permanent charges are created by the isomorphic substitution of cations in the lattice structures of clay mineral. Substitution of a lower valency cation (Mg2+) with a higher valency cation (Al3+) creates net positive charge and the substitution of a higher valency cation (Si4+) with a lower valency cation (Al3+) creates a negative charge. The variable charges are created through the dissociation of H+ and OH– ions from the edges of the clay mineral and from the functional groups of organic matter. The major variable charge components in soils include organic matter, iron and aluminium oxides, kaolinite and allophane.

In variable charge soils the surface charge varies with pH; the negative charges increase with an increase in soil pH and the positive charges increase with a decrease in pH. In these soils there will be a pH at which the soils carry equal number of positive and negative charge and the net charge becomes zero. This pH is known as point of zero net charge (PZNC). In variable charge soils when the soil pH is below its PZNC then the soil is net positively charged and when the pH is above PZNC the soil is net negatively charged. The PZNC of a soils depends on the nature of the variable charge components. Soils dominated by iron and aluminium oxides and allophane have a high PZNC and soils dominated by organic matter is less than the field soil pH. Hence these soils are net negatively charged and adsorb insignificant amounts of non-specifically adsorbed anions, such as nitrate and chloride. Whereas the PZNC of some of the subsurface soils which are dominated by allophane is more than the field soil pH. These soil are net positively charged and adsorb significant amount of nitrate and chloride through non-specific adsorption. Therefore, nitrate and chloride can be considered as reactive ions in soils containing a significant amount of positive surface charge.

Biological Transformation

The mobility of anions not only depends on the extent of adsorption onto soil particles but also on the extent of biological transformation. In general, nitrate, sulphate and phosphate are biologically more reactive than chloride.

When plant residues with high carbon content are added to soils the microorganisms tend to use some of the mineral nitrate and sulphate. These inorganic mineral forms which are available for plant uptake and leaching are converted to organic form (immobilisation). Immobilisation process reduces the leaching of nitrate, sulphate and phosphate.

Pesticides

The main process affecting the efficiency, and ultimate fate of pesticides in the soil are retention (sorption-desorption), transformation (biological and chemical degradation), plant uptake and transport into the soil, the atmosphere and surface waters. These processes are more or less interdependent. Sorption, for example, may protect pesticides against microbial attack but also facilitate their abiotic transformation (Bollag and Liu, 1990). By inhibiting desorption (release) into the soil solution, sorption also retards pesticide transport into the soil and surface run-off (Weed and Weber, 1974).

Mechanism of Pesticide Interaction in Soil

There is increasing concern about the persistence of pesticide residues in soils and subsequent contamination of surface and groundwater. The persistence of pesticide in soils largely depends on the extent of adsorption onto soil particles and the rate of mineralisation or degradation (Necholls, 1988; Alexander, 1994). While adsorption decreases the leaching of pesticide residues to groundwater, erosion of pesticide-enriched finger soil particles by runoff would lead to surface water contamination (Wauchope, 1978). Degradation is one of the main pathways by which pesticides are removed from soils (Alexander, 1994). An understanding of the degradation of the toxic organic pesticides into less toxic or non-toxic compounds is essential for predicting pesticide persistence in the environment and developing strategies for risk management. Soil microorganisms generally play the major role in the degradation of organic pesticides.

In soils, both adsorption of organic pesticides and the biological activity increase with an increase in the organic matter content (Johnson and Sims, 1993). These two factors have opposite effects on the degradation of pesticides. Adsorption has often been shown to decrease the degradation rate of organic chemicals by reducing their availability to microbial attack (Ogram *et al.*, 1985; Smith *et al.,* 1992). An increase in soil microbial activity is likely to increase the rate of degradation pesticides. The influence of microbial activity on pesticide degradation, however, has often been inferred indirectly from the effect of other factors such as temperature, moisture and nutrient supply on degradation (Alvey and Crowley, 1995).

Pesticide interaction in soils vary considerably depending on the nature of soil and chemicals. Depending on the surface properties of the sorbents and the charge characteristics of the sorbate, different sorptive mechanisms are thought to be operative. Of which, ion exchange, cation-bridging and ion-dipole interactions are important (Bailey and White, 1970; Mortland 1970; Theng, 1974). In addition, ligand exchange, charge transfer, hydrogen bonding, and Van der Waals interactions may also come into play. The interactions of non-ionic pesticides are largely through the soil organic phase. It has been reported that simple partitioning of the chemicals between the aqueous and the organic phase often predominates over site interactions (Chiou *et al.*, 1979). On the other hand, cationic pesticides such as diquat and paraquat are principally sorbed by cation exchange, that is, by replacing inorganic cations present on clay and organic matter surfaces (Kookana and Aylmore, 1993). Retention is by electrostatic (columbic) attraction to the negatively charged sites on clay and organic matter surfaces and are strongly retained by soil. However, the energy of interaction is apprantly higher for clay than for organic matter. Ionic pesticides interact weakly with negatively charged surfaces of clay and organic matter and increases with a decrease in pH (Baskaran *et al.*, 1996).

Pesticide Adsorption

The adsorption isotherms for pesticides are often adequately described by the Freundlich equation (1).

$$S = K_f CN \qquad \text{.......(1)}$$

where S is the amount of pesticide adsorbed (mg/kg), C is the equilibrium (mg/dm^3) in solution and K_f and N are adsorption coefficients. When N=1, equation 1 becomes linear (2):

$$S = K_d C \qquad \text{.......(2)}$$

where K_d is the distribution coefficient (dm^3/kg). N values close to 1 have been reported for pesticide adsorption in many studies (Singh *et al.* 1989; Mallawatantri and Mulla, 1992) indicating that the adsorption reaches linear form. Simple correlation analysis between pesticide adsorption, as measured by the distribution coefficient (K_d), and various soil properties (pH, CEC, organic carbon, sand, silt, clay, and oxalate extractable Fe and Al) has indicated that often the organic carbon content gives a significant correlation with the K_d values. The K_d values generally increased with an increase in the organic matter content of the soils. Baskaran *et al.* (1996) observed that allophanic soils with a high organic matter content adsorbed greater amounts of organic pesticides than do non-allophanic soils with a relatively low organic matter content.

Since organic carbon (OC) is considered to be the principal soil component involved in pesticide sorption, the K_d values are generally normalised to the organic carbon content of the soils (3):

$$K_{oc} = (K_d / \% \, OC) * 100 \qquad \text{.......(3)}$$

where K_{oc} is the distribution coefficient based on soil organic carbon. It has often been observed that although the K_d values of organic pesticides increase with organic matter content of soils, the K_{oc} values varied less between soils. Based on this the K_{oc} value of a pesticide is considered to be an unique property and these values are often used in conjunction with the half life ($t_{½}$) values in predicting the leaching potential of pesticides (Jury *et al.*, 1986). The K_{oc} values of ionic pesticides often found to be higher at low levels of organic carbon indicating that clay also contributes to adsorption in these soils. Baskaran *et al.*, (1996) observed an increase in K_{oc} values for 2,4-D with soil depth which they attributed to the greater contribution of clay mineral to 2,4-D, adsorption in the lower layers than in the surface layers. While the non-ionic form is mainly retained onto the soil organic matter the ionic form is retained onto both the organic matter and the clay minerals.

Pesticide Desorption

The amount of pesticide desorbed is often found to decrease with an increase in soil organic carbon. It has been observed that when adsorption occurs mainly on silicate clays near complete desorption of the pesticide is obtained whereas when the adsorption occurs mainly on the organic matter only partial desorption is obtained (Baskaran, 1994).

The concentration of pesticides remaining in the adsorbed phase decreases with increasing desorption time and a curvilinear relationship is obtained between the cumulative time and the amount of pesticide remaining in the soil. This relationship followed a first-order reaction kinetics with respect to surface concentration (equation 4)

$$S_t = S_i \exp(-K_1 t) \qquad (4)$$

where S_i and S_t are the amount of pesticide (mg/kg) in the soil phase at time 0 and t, respectively, t is desorption period (min) and K_1 is the rate of desorption (1/min).

It has often been observed that for short duration the desorption of pesticide follows first-order reaction kinetics (Smith *et al.* 1992). The rate of desorption decreased with an increase in the organic carbon content of the soils. An inverse relationship between rate of release of pesticides and organic matter has also been observed in solute transport studies (Brusseau and Rao, 1989). This inverse relationship suggests that the desorption of pesticides involves the diffusion of the compounds through soil organic matter.

Pesticide Degradation

The rate of degradation of pesticides is measured from the amount of pesticide remaining in the soil or from the amount of CO2 released. The rate of degradation is often found to follow the first-order reaction kinetics (equation 5) (Roeth et al. 1990; Campbell et al.1991).

$$C_t = C_i \exp(-k_2 t) \qquad (5)$$

where C_i and C_t are the amount of pesticide in the soil at time 0 and t, respectively, t is the degradation period (day) and k_2 is the rate constant (1/day).

The time (day) taken to reduce the pesticide to half of its initial concentration (half life; $t/_{½}$) is estimated from equation (6):

$$t^{½} = 0.693/k_2 \qquad (6)$$

The $t_{½}$ values for organic pesticides in soils vary depending on the nature of the organic compound and the soil and climatic conditions.

Bolan and Baskaran (1996) have observed that the rate of degradation of 2,4-D, as measured by the $t_{½}$ value, increased with an initial increase in K_d values, but it decreased when the K_d values were greater than 8 dm³/kg. The initial increase in $t_{½}$ values (decrease in the rate of degradation) was probably caused by the adsorption of the pesticides on to soil. This could result in its inaccessibility to microbial attack (Smith *et al.*, 1992) and a low concentration of the pesticide in solution. Processes which lead to a decrease in the concentration of the pesticide in soil solution are likely to decrease the rate of degradation of the pesticides by the microorganisms (Alexander, 1994).

There is evidence in literature that the rate of desorption decrease with increasing adsorption capacity and hence the rate of desorption becomes the limiting factor for microbial degradation (Felsot and Dahm 1979; Rijnaarts *et al.* 1990). Adsorbed pesticides becomes "bound" or "recalcitrant" which resist degradation and become more persistent in soils (Fewson 1988). Using mechanistic models, Ogram *et al.* (1985) showed conclusively that adsorbed 2,4-D was completely protected from biological degradation. They suggested that adsorbed 2,4-D is located sufficiently deep within the soil-organic matter matrix that the bacteria are unable to attack it. The adsorbed 2,4-D needs to be desorbed into soil solution for it to undergo microbial degradation. The desorbed 2,4-D is likely to undergo microbial degradation. Thus under the field conditions, depending on the rate of desorption, the adsorption of 2,4-D would provide a temporary protection from degradation.

Soils with high K_d values tend to have high organic matter content and mircobial activity. The decrease in the concentration of pesticide compounds are important in controlling the rate of degradation. Lehr *et al.* (1996) observed that microbial biomass correlated with the mineralisation of free isoproturon and not with cell wall bound isoproturon. This is attributed to the lower bioavailability of cell bound compound resulting in a higher persistence to microbial degradation. Binding of isoproturon to plant cell walls may additionally impair the free access of microorganisms to the pesticides. Similarly Alvey and Crowley (1995) observed that the addition of rice hulls to soils enhanced atrazine degradatiion which they attributed to the improvement in the survival and activity of degrader consortium. Also a decrease in the rate of degradation with increasing soil depths has often been attributed to a decrease in the microbial population (Campbell *et al.*, 1991). The nature of contaminant interactions and the extent to which they bind to colloid particles control their impact on soil quality. The adverse impacts of contaminants are briefly discussed in the following sections.

FERTILIZER POLLUTION

Increased agricultural productivity through improved technologies has helped in maintaining the balance between food supply and demand. The supply of nutrients from the soil from recycled organic matter and biological fixation of N is not sufficient to ensure the present level of food production. Inorganic fertilizers are, therefore, used to provide additional plant nutrients to ensure crops produce their optimal yields while maintaining the fertility of the soil. According to an estimate, fertilizers accounted for 75 per cent Asian cereal growth during 1979-89 (Anonymous, 1992a). Chemical fertilizers, no doubt, constitute an invaluable input for producing required food crops.

With increasing concern among policy makers, scientists and the general public, the intensive high input farming is believed to be the cause of considerable damage to environment as a result of air, water and soil pollution leading to loss of genetic diversity and ecosystem disturbance. Excessive use of fertilizers is one of the important parameters in question. Several problems linking excessive fertilizer use with environment have been identified. Increase of nitrates in drinking water, something about which general public feels greatly concerned and believed to be due to excessive use of N fertilizers and animal manures, seems to be the most important fertilizer related pollution issue. Nutrient enrichment, eutrophication and deterioration of surface water quality due to transportation of nutrients applied through fertilizers via leaching and/or runoff and sediment erosion are other problems. Contamination of soils by heavy metals through fertilizers such as cadmium from phosphatic fertilizers is also getting increasing attention of environmentalists (Kostial, 1986). The significance of fertilizers in polluting the air is being viewed with concern after reports that nitrous oxide originating from agricultural soils can damage the ozone layer in the stratosphere (Bouwman, 1990; Smith *et al.*, 1990). Fertilization as an element of intensive agriculture is also partly responsible for the impact of modern farming activities on the environment in terms of impairment, reduction in size, splitting up and destruction of natural and semi–natural biotopes and rural landscape elements.

In the developed countries, there has been intensive fertilizer use for the last four decades. If the polluting effects of fertilizers are being observed now, similar problems in developing countries like India should be expected in the near future. In India, pollution effects associated with fertilizers should occur first of all in the Punjab where fertilizer application rates are the highest in the country and soils are intensively irrigated.

Fertilizer consumption has increased greatly all over the world in past forty years. Worldwide, total application of N through fertilizers was around 60 million tonnes in 1981-82 and it has increased to more than 77 million tonnes in 1991-92. Thus, as developing countries like India strive to produce more food by applying large amounts of plant nutrients through fertilizers, it becomes increasingly important to consider seriously the fertilizer related environmental problems. Consumption of NPK in India during 1993-94 was 12.3 million tonnes and that in the states of U.P. and Punjab, it was 2.3 and 1.2 million tonnes respectively (FAI, 1994). Since 10 per cent or more of total nutrients applied through fertilizers in India is consumed in Punjab having only around 3 per cent of thc nct cultivated area of the country, the effect of fertilizer pollution shall be felt in Punjab quite early.

ADVERSE IMPACT OF AGROCHEMICALS ON SOIL QUALITY

There are a number of ways in which agricultural practices can impact on soil quality. Only some of the important issues are discussed below. use of both nitrogenous and phosphate fertilizers can impact on soil acidity. A large proportion of acidic soils in Australia have arisen from indiscriminate use of such fertilizers.

In legume-based rotation, nitrogen (N) is derived mainly from the biological fixation of the atmospheric N by a group of bacteria (known as rhizobium) living in the root nodules plants. A small amount of N fertilizer is applied to pastures during the winter season when the rate of N fixation is slow. Recently there has been a sudden increase in the use of N fertilizers in the dairy pasture. The excessive use of N fertilizers has rekindled the debate on the acidifying effects of N fertilizers added to soils.

Traditionally, most phosphorus (P) has been applied to agricultural land in the form of water soluble fertilizers, such as single superphosphate, triple superphosphate and ammonium phosphates. In recent years there has been renewed interest in the use of slow-release phosphate fertilizers which include phosphate rocks (PRs) for direct application and partially acidulated phosphate rock fertilizers (Bolan *et al.*, 1990a). There is currently a debate on the acidifying effects of different phosphate fertilizers added to soils. For example it has been claimed that the use of superphosphate fertilizers in legume based pasture causes soil acidification, whereas the direct application of reactive phosphate rocks 9RPR) as a source of phosphorus decreases the rate of soil acidification (Sinclair *et al.*, 1993).

NITROGEN FERTILIZERS

Plants take up N in three main forms - as an anion (nitrate, NO3–), as a cation (ammonium, NH4+) or as a neutral N2 molecule (N2 fixation). Depending upon the form of N taken up and the mechanism of assimilation in the plant, excess of cation or anion uptake may occur. To maintain the charge balance during the uptake process, hydrogen (H+) or hydroxyl (OH–) ions must pass out of the root into the surrounding medium. The H+ ions may be derived from the dissociation of organic acids within the plant cell and the OH– ions from the decarboxylation of organic acid anion. It has been shown that in general, while the uptake of NH4+ and N2 fixation results in a net release of H+ ions, uptake of NO3– can result in a net release of OH– ions. It is to be pointed out that the net release of H+ ions results in the acidification and the net release of OH– ions results in the alkalinisation.

Application of N fertilizers, such as urea and ammonium sulphate to soils produces acid by two processes. Firstly the natural process of oxidation of ammonia ions to nitrate ions (nitrification) releases acid. Part of the acid produced is neutralized by alkaline ions released by the plants during the subsequent uptake of the nitrate ions. Secondly since nitrate ions are not strongly adsorbed by the soil they are liable to leach or move down through the soil. The negatively charged nitrate ions carry positively charged basic cations, such as Ca, K, Mg and Na in order to maintain the electric charge on the soil particles. It is the depletion of these basic cations during the leaching of nitrate ions which accelerates the acidification process. In the case where ammonium sulphate is added to soils the movement of both the movement of both the sulphate and the nitrate ions in the soil causes greater depletion of basic cations. In the case of urea, the initial conversion of amide nitrogen to ammoniacal nitrogen (ammonification) releases alkaline ions which neutralise part of the acid produced during the subsequent oxidation of ammonium ions to nitrate ions. This is one of the reasons why urea-based N fertilizers are less acidifying than the ammonium based fertilizers. It is estimated that approximately 43, 102 and 131 kg lime (calcium carbonate; $CaCO_3$) is required to overcome the acidity produced by the application of 25 kg N per hectare as urea, diammonium phosphate and ammonium sulphate, respectively.

The extent of acidification caused by N fertilizers depends mainly on the magnitude of leaching of the anions and the basic cations present in the fertilizers. Since phosphate ions are less liable for leaching than other anions, such as nitrate, chloride and sulphate, N fertilizers which contain phosphate as anions (*e.g.*, ammonium phosphates) cause less acidification than those containing the other anions. Similarly N fertilizers containing basic cations, such as calcium nitrate and potassium nitrate are likely to cause soil acidification.

In legume-based cropping systems, fixation of atmospheric N and the subsequent leaching of nitrate formed from the mineralization of the fixed N result in the acidification of the soils. It has been shown that approximately 400-550 kg $CaCO_3$/ha is required annually to neutralise the acidity produced by these processes under these systems.

PHOSPHATIC FERTILIZERS

In legume based cropping systems, phosphate fertilizers produce acids by two processes, namely (i) fertilizer dissolution (direct) and (ii) nitrogen fixation (indirect).

Monocalcium phosphate (MCP) is the principal phosphate component present in superphosphate fertilizers. Dissolution of MCP in soils results in the formation of dicalcium phosphate (DCP) with a release of phosphoric acid (Equation 7) close to the fertilizer granules (Lindsay and Stephenson, 1959).

$$Ca(H_2PO_4)_2.H_2O + H_2O \rightarrow CaHPO_4.2H_2O + H_3PO_4 \quad \text{.......................(7)}$$

Phosphoric acid subsequently dissociates into phosphate ions and hydrogen ions (acidity). Part of this acidity is subsequently neutralised by the alkaline ions released during the adsorption of the phosphate ions by soil particles. Unlike the nitrate ions since the phosphate ions are strongly adsorbed by most soils, phosphate induced leaching and depletion of basic cations is unlikely to occur.

In legume based cropping systems, phosphate fertilizers are added mainly to promote the N fixation by the legumes by overcoming a soil deficiency of P. Fixation of atmospheric N and the subsequent leaching of nitrate formed from the mineralization of the fixed N result in the acidification of the soils (Bolan *et al.*, 1990b). It is important to point out that irrespective of the P fertilizer sources, application of P to legume based cropping systems, promotes N fixation and thereby indirectly causes soil acidification. The amount of acidity produced indirectly by N fixation depends mainly on the extent of nitrate leaching and is much higher and affects the bulk of the soil far more than the acidity produced directly by the dissolution of MCP in superphosphate fertilizer granules.

LIMING VALUE OF ROCK PHOSPHATE

Increasing quantity of phosphate rocks (PRs), are added directly to soils as a source of phosphorus. In addition to supplying phosphorus and calcium, PRs can also have a liming value (Loganathan et al., 1994; Manoharan et al., 1994). The liming action of PRs can occur through two processes.

Firstly, most PRs contain some free calcium carbonate ($CaCO_3$) which can act as a liming agent in its own right. Secondly the dissolution process of the phosphate mineral component (known as apatite) in soils consumes acids (Eq 8) and thereby reduces the soil acidity (Bolan and Hedley 1989).

$$Ca_{10}(PO_4)_6F_2 + 12H^+ \rightarrow 10Ca^{2+} + 6PO_4^- + 2F^- \quad \text{.............(8)}$$

POTASSIUM FERTILIZERS

Potassium fertilizers generally do not have large effects on soil acidification. High application rates of potassium fertilizers, such as potassium chloride (KCl) and potassium sulphate (K2SO4) result in the leaching of the chloride and sulphate anions which can deplete calcium ions. Calcium losses may be high but if there is no net loss of basic cations (K substituting Ca) the soil will not become more acid.

SULPHATE FERTILIZERS

Gypsum, a component of single superphosphate and elemental sulphur are the most common sources of sulphur. Since gypsum contains both calcium and sulphate ions, leaching of sulphate is unlikely to deplete completely the added calcium. When elemental sulphur is added to soils it is oxidised to sulphuric acid which dissociates into sulphate ions and acidic hydrogen ions.

$$2S^{\circ} + 3O_2 + 2H_2O \rightarrow 2H_2SO_4 \quad \text{..........} \quad (9)$$

Unless the the sulphate produced is lost through leaching, part of the acid produced is neutralised by the alkaline ions released during the adsorption of sulphate ions by soil particles and immobilization of sulphate ions by soil microorganisms. It is estimated that a maximum of approximately 3.1 kg $CaCO_3$ is required to neutralise the acidity produced from the oxidation of 1 kg of elemental sulphur. It is to be remembered that sulphur is being added to overcome a deficiency of S in soils to enable legumes to maximise N fixation as is the case with adding P to overcome P deficiency. Therefore, the indirect effect of S addition on N fixation is more important in creating acidity in soils, than from the oxidation of elemental S.

Nitrate Pollution

Nitrogen is added to the soil through nitrogenous fertilizes, legumes, animal manures, crop residues and other wastes. Depending on the rate and method of application of fertilizer N, the kind of crop and soil and climatic variables, 35 to 60 per cent of the applied N is usually recovered by crop plants; 10 to 20 per cent may be converted to N2 gas, nitrous oxides or ammonia and volatilize into the atmosphere; and if there is excess water percolating through the soil, part of the remaining N may move below the root zone or into the ground water.

Leaching of applied nitrate-N depends upon amount of water percolating through the soil and the presence of excess nitrate-N in the soil. According to Bijay-Singh and Sekhon, 1979, situations favouring to nitrate leaching include periods when rainfall exceeds evapotranspiration or where irrigation is practiced, lands devoted to shallow rooted crops, soils having low water holding capacities and high infiltration rates, and where fertilizers and manures are applied in amounts and ways that result in the presence of excess amounts of nitrate-N in the soil.

Although there has been a concurrent rise in fertilizer usage with nitrate concentration in water, particularly where irrigation is practised, there are many other sources of nitrate-N namely sewage effluent, animal excreta, natural soil nitrate and decomposition of soil organic matter (Keeney, 1989). In many regions, natural deposits of nitrate may constitute a major source for nitrate pollution of groundwater. In selected locations in Punjab, Haryana and Western Uttar Pradesh, nitrate-N concentration (Table 4.2) in well waters samples was several fold higher than the upper safe limit (Handa, 1987).

Carcinogenic nitrosamines by reaction of ingested amines with nitrite in the human stomach may present a further health hazard after exposure to high levels of nitrate. A large number of N-nitroso compounds have been shown to induce tumors in test animals in many tissues (Brezonik, 1978). Data relating exposure to nitrate or nitrite to the incidence of cancer have also been reported (Fraser and Chilvers, 1981). Nitrate-N may also affect the cardiac functions of man.

TABLE 4.2: Nitrate content of well waters

District	State	NO_3^--N (mg/l)
Aligarh	Uttar Pradesh	61
Meerut	Uttar Pradesh	157
Mahendragarh	Haryana	296
Ambala	Haryana	223
Bhatinda	Punjab	128
Sangrur	Punjab	98

Source: Handa (1987)

Heavy Metals

Health hazards associated with heavy metals entering the food chain through soil is demanding attention. Fertilizers contain heavy metals as impurities. The application of rock phosphate or its products to soil always implies the addition of a significant amount of lead and cadmium into the soil. Analysis of several commercial fertilizers (Table 4.3) commonly used in Punjab revealed that a combination of low analysis and straight fertilizers can add more lead and cadmium to soil than high analysis and mixed fertilizers (Arora et al., 1975; Bijay-Singh and Sekhon, 1977b).

TABLE 4.3: Lead and cadmium content (mg/kg) of fertilizers

Fertilizer	Lead		Cadmium
	(a)	(b)	(a)
Urea	-	4	1
Calcium ammonium nitrate	116	200	6
Muriate of potash	117	88	14
Diammonium phosphate	195	188	109
Superphosphate	487	609	187
Rock phosphate	962	1135	303

Source: (a) Arora et al. (1975); (b) Bijay-Singh and Sekhon (1977b)

Several reports suggest that heavy metals applied to the soil through different derivatives of rock phosphate accumulate almost completely in the surface layer of the soil and in forms generally easily available to plants (Williams and David, 1973). In coarse textured soils with low pH, availability of heavy metals applied through fertilizers was more than in those containing large amount of clay and with alkaline reactions (Mortvedt and Osborn, 1982; Rana and Kansal, 1983). Average annual cadmium deposits in European soils which can be attributed to fertilizers are less than 0.5 per cent of the amount already existing in these soils.

Eutrophication

Eutrophication is the process of enrichment of surface waterbodies like lakes, reservoirs and streams with nutrients. Nutrient enrichment of waterbodies results in intense prolification and accumulation of algae and higher aquatic plants in excessive quantities which can result in detrimental changes in water quality and can significantly interfere with use of the water resource.

Surface waters also receive largest contribution of nutrients from non-point sources such as agricultural lands. Several studies have demonstrated that high levels of total N in surface waters are generally related to farming activities. A large part of the nutrient load generated from agricultural land is related to the natural nutrient content of the soil. Practices which do not allow for the proper incorporation of fertilizers applied at excessive rates can cause short term elevated nutrient concentrations in runoff (Bijay-Singh and Dhaliwal, 1993).

Surface runoff on sloping land can also influence the quality of waters through colloidal transport. Coloidal particles can act as vehicles of contaminant transfer. Indeed, algal bloom in the River Murray basin in Australia has been attributed to surface transport of nutrients bound to soil colloid particles. During algal bloom certain poisonous chemicals are released. Such poisonous chemicals release has killed fish and other river life and this has been documented in the news many times in Australia. Toxic effect of these chemicals to swimmers have also been recorded. For example, algal blooms have had an adverse effect on salmon and trout farms in Norway; this damage has incurred costs in excess of about $ 200 million. Inspite of this problem European countries continue to dump 1.5 million tonnes of N into adjacent marine areas, more than 60% of which is derived from agricultural runoff (Saull, 1990).

In USA more than 50% of the community and rural wells are contaminated with NO_3–N with 2% exceeding the drinking water standard of 10 mg NO_3^-N/l. Nitrate in combination with dissolved organics can produce nitrosamines that are highly carcinogenic. In addition, intensive agricultural production also requires high amount of nitrate fertilizers (N > 200 kg/ha) which creates an increased potential for NO_3^-–N leaching to groundwater.

Stratospheric Changes

Losses of N from the soil-plant system takes place by way of ammonia volatilization, denitrification and chemodenitrification. Loss of applied fertilizer N takes place as NH3 (ammonia), N2O (nitrous oxide), N2 (dinitrogen), NO (nitric oxide) and NO2 (nitrogen dioxide). Dissimilatory denitrification is carried out by a limited number of aerobic bacteria that can grow in the soil in the absence of molecular oxygen while reducing nitrates and nitrites to N2O and N2. Chemodenitrification is a term that encompasses processes for gaseous losses of N from soils through chemical reaction of nitrite.

Many research studies have been involved in view of the concern that N_2O released into the atmosphere may increase the rate of reactions in the stratosphere that lead to the destruction of ozone layer. The stratosphere ozone layer shields the biosphere from the harmful ultra violet (UV) radiation and also influences the earth's temperature. According to Crutzen (1983) N_2O from N-fertilizers does not have a serious impact on ozone content of stratosphere. Ammonia containing fertilizers favour N_2O emission and concentration. Katyal *et al* (1989) reported that organic manure and application of anoxic soil conditions depress N_2O levels. As per the estimate given by Weiss (1981), the global increase in atmospheric concentration of N_2O is about 0.2 per cent per year which may be attributed to 3.5 per cent per year increase in N_2O emission from fossil fuel combustion.

Fertilizer Use Efficiency

Fertilizer use efficiency is the per cent recovery of fertilizer N by a crop. It varies with the type of fertilizer and rate at which N is applied, the nature of the chemical and biochemical reactions between soil and fertilizer, the timing and placement of fertilizer, the type of crop and its N requirement, the adequacy of other nutrients and a host of soil, climatic and management factors. Under favourable conditions, 80 per cent or more of the fertilizer N may be recovered by the crop but under most situations efficiencies of 50 per cent or less are common. Low fertilizer use efficiency is caused by loss of N from the soil plant system through mechanisms such as leaching, ammonia volatilization and denitrification. Thus, fertilizer use efficiency is an important consideration if higher levels of crop yields are to be attained and pollution of the environment with plant nutrients is to be prevented. According to Follett (1992), the two general approaches for minimizing fertilizer related pollution are (i) optimize use of the crop's ability to compete with processes whereby plant available N is lost to the environment, and (ii) direct lowering of the rate, duration and amount of loss caused by loss processes themselves.

Bijay-Singh *et al.* (1978) described an approach to determine optimum rates of fertilizer N at which yields are least affected while unused N, which is potential pollution threat, is reduced to a permissible level. Functions relating yield and N uptake with applied N are used to define fertilizer N application rates (i) at which greatest economic returns are obtained, and (ii) at which the amount of applied N in excess of the environmentally permissible amount of N in the soil profile is absorbed by the crop. To compute the latter quantity, the permissible level of un–utilized N must be known. The optimum rate is the smaller of the two above defined fertilizer N rates. The basic philosophy of these approaches is to apply fertilizer N at rates to ensure high fertilizer use efficiency so that the amount of un–utilized N is reduced to environmentally acceptable levels.

There is no realistic alternative but to use increasing quantities of organic fertilizers for maintaining food supplies in order to feed the increasing population. Used properly, fertilizers can, in fact, lead to better environment because greater vegetative growth and higher crop yields (i) leave large quantities of residues in the soil, thus reducing erosion, (ii) absorb more carbon dioxide, a major source of green house effect, and (iii) lessen the need to bring marginal and forest land into cultivation. Although application of rock phosphate or its products always result in the addition of different heavy metals into the soil; fertilizers, in general, do not constitute a significant source. The impact of N_2O production due to increasing use of N fertilizers in damaging the stratospheric ozone layer is also unlikely to be great.

ENVIRONMENTAL EFFECT OF SOIL ACIDIFICATION

Soil acidification has several consequences as far as soil quality is concerned. First, acidic pH can impact on soil microflora and this can in turn influence the carbon dynamics of the soil environment. Second and most importantly, acidification can produce elevated levels of soluble Al3+ and Mn2+ (Equations 10 & 11) which are highly toxic to many crops and plants.

$$Al(OH)_3 + 3H^+ \rightarrow Al_3^+ + 3H_2O \quad (10)$$

$$MnO_2 + 4H^+ + e^- \rightarrow Mn_2^+ + 2H_2O \quad (11)$$

Loss in soil quality due to elevated metal ions (Equations 5 & 6) leads to decreased crop yield due to decreased pH, loss of basic cations (e. g., Ca and Mg), inhibition of nitrification, promotion of certain diseases and fixation of certain major anionic nutrients.

In addition to impacting on soil quality, use of superphosphate also introduces toxic heavy metals such as Cd, Pb and Hg in the soil environment. Cadmium is present in the parent rock phosphate that is used for the manufacture of superphosphate. Once present in the soil environment, Cd being highly reactive accumulates in the surface soil environment that plant roots proliferate. Therefore, this toxic metal can gradually transfer into crops thus impacting on their quality. Presence of Cd in consumables has implications to animal and human health and it can also affect the marketability of our produce.

PESTICIDAL POLLUTION

During the traditional technology period in India, the use of chemical pesticides were limited mainly to the high value crops only. The traditional varieties did not face much problem of insect pests and diseases and if at all they occurred, the traditional methods available could control them. However, with the advent of the high yielding varieties and adoption of intensive cropping practices, brought to the forefront pest problems which were quite peculiar and demanded high pesticide use to check the yield losses. Pesticide consumption has shown a steady increase. Plant protection coverage increased from 2-4 million hectares in 1956 to 80 million hectares in 1984 and the quantity of pesticide rose from 200 tonnes in 1955 to 72,000 tonnes in 1987 (Table 4.4). However, the use of these poisonous substances, although essential, needs lot of caution and adoption of safety measures. Pesticides use is contributing towards imbalances in the ecosystem and polluting the environment. Thus, safe use of pesticides should be given priority.

TABLE 4.4: Pesticide consumption in India

Year	Consumption (00'kg)
1980	292084
1985	493990
1986	480552
1987	719906

Source: FAO Production Year Book (1988)

Pesticides are now extensively used to control various pests which are harmful to the crops raised by man for food, feed and fibre production. They have played an important role in boosting agricultural production during the past four decades. However, most of the chlorinated pesticides are non-biodegradable and leave residues which are detrimental to human beings, animals and the environment. Some of the harmful pesticides have either been banned or restricted specially in developed nations. Residue studies in India have also revealed wide spread contamination of the food products and the environment as a whole.

Insecticides

The presence of residues of insecticides in food commodities and other components of the environment is a matter of serious concern. Even small quantities of the residues ingested daily alongwith food can build up to high levels in body fat (Dhaliwal and Singh, 1993). There is, therefore, a constant fear that society is being slowly poisoned by the intake of food contaminated with pesticide residues.

During the 1900's, there was little information available on the impact of pesticide on soil and water quality. Consequently, pesticides in excess of doses which are now considered safe were applied to the soil environment. Organochlorine insecticides such as, chlordane, dieldrin, DDT and heptachlor were widely used in agriculture/horticulture to control insect pests in Australia and New Zealand in 1960's. Since then its use was progressively restricted, and finally banned now. Nevertheless, their residues still found in soil, and continue to cause problems of food and feed contaminants. Chlordane, DDT, dieldrin and heptachlor were detected in 18.4, 39.6, 39.0 and 18.38 %, respectively of 11248 samples from Western Australia (EPAWA 1989). In New Zealand, export beef contain>1 mg/kg DDT residues (MacIntyre *et al* 1989). The metabolite of DDT (DDE) still occur in the top 150 mm of the soil profile in Canterbury region in New Zealand (Boul 1994).

There are thousand of arsenic/DDT contaminated soils through out Australia. It has been estimated that there are about 1700 cattle, and over 60000 sheep dip sites in North Eastern New South Wales alone. Concentrations of DDT exceed 10% at some sites and this has resulted in total elimination of soil micro–ecosystem (Beard 1993). DDT per se is not considered toxic but in the presence of conditions that are prevalent in moist or reducing soil environment DDT degrades to form toxic daughter compounds DDD and DDE. Under anaerobic conditions, DDT is degraded to DDD and to a lesser extent to DDE. In the presence of O_2, DDT is degraded to DDE and then to DDD. These breakdown products are highly toxic to soil microorganisms, animal and human beings. These pesticides were later replaced by organophosphorus compounds such as parathion and malathion which are many times more toxic, particularly to the person spraying them, than DDT. There are many organophosphorus contaminated sites in Victoria and New South Wales, Australia. Other pesticides (*e.g.* organo sulfur compounds) have also been used indiscriminately and these also impact on the quality of the environment as some of these have been documented in surface and ground waters. Even after more than 80 years of pesticide usage, the chemical reactions involving pesticides in general use in the soil environment are still virtually unknown.

In case of wheat grains in India, main contaminants have been DDT, HCH and to some extent other organochlorine insecticides such as aldrin and heptachlor. A few samples from Ludhiana and Mysore were found to contain malathion ranging from 0.02 to 11.92 ppm which appears to be a case of direct spray of malathion on the gunny bags/grains in the storage godowns. The samples analysed during 1960s and 1970s contained relatively more amount of DDT (Tr-9 ppm), HCH (Tr-13.9 ppm) whereas samples analysed after 1980 showed contamination level to be much lower than earlier studies (DDT, Tr-1.75 ppm; HCH, Tr-3.92 ppm). The seasonal vegetables consisting of green leafy vegetables, tomato, potato, cauliflower, okra, cabbage, etc., have been analysed mainly for residue content of organochlorine insecticides such as DDT, HCH, aldrin, dieldrin, heptachlor and lindane. Most of the samples have been found to be contaminated with DDT, HCH or both. However, a few of them showed the presence of aldrin, dieldrin, endrin, heptachlor and endosulfan (Kathpal and Kumari, 1993).

The contamination of bovine milk with residues of HCH has been reported to be quite alarming. About 90 per cent of 980 samples of milk samples analysed in the states of Andhra Pradesh, Maharashtra, Delhi and Punjab showed HCH residues at levels ranging from 0.12 to 40 ppm on fat basis (Kalra and Chawla, 1981). Samples analysed under the ICMR project also reveal that there is a widespread contamination (81% samples) with HCH residues (ICMR, 1993). Human breast milk is also known to contain residues of insecticides, most commonly those of DDT and HCH. The mean level of total DDT, HCH and aldrin was 0.52, 0.2, 0.04 mg/kg in 50 samples collected from rural areas in two districts of Punjab (India) also showed high levels of residues of DDT and HCH (Singh and Dhaliwal, 1993). The milk samples from Lucknow were contaminated with aldrin, the cleavage level being 0.02 ppm in buffalo and 0.005 ppm in goat milk (Saxena and Siddiqui, 1982). Contamination of other materials like, ghee, cheese, butter, honey etc. has also been reported by several workers.

Fungicides

Agriculture is now being transformed into dynamic productive and profitable sector because of the ever increasing demand for food and feed products. In order to cope up with the food production targets, more emphasis is likely to be given to agricultural inputs. By the end of the century, the average agricultural production in the developing countries shall have to be increased by 30% over the present level. The use of chemicals as plant protection practices acts as an insurance against the risks of loosing not only the crop yield but also the money spent on advanced technologies used as inputs in attaining higher production. Breakthroughs in the development of different agrochemicals and their widespread use at the farm level indeed have helped to boost the agricultural production in all countries. Fungicides constitute 10 per cent of the total consumption of pesticides in India and 17 per cent in the world (Table 4.5).

TABLE 4.5: Pesticides Market Share

Pesticides	Market Share (%)	
	World	India
Herbicides	45	7
Insecticides	36	80
Fungicides	17	10
Others	2	3

Source: Sachan (1989)

TABLE 4.6: Global crop/fungicide matrix

Crop	Fungicides (million dollars)
Fruits and Vegetables	1300
Rice	380
Wheat	345
Soybean	50
Sugarbeet	45
Cotton	40
Maize	40

Source: Sachan (1989)

During the previous ten years period, the plant protection coverage in India increased from approximately 8 m ha to 14 m ha, acquiring an increase of 75 per cent (Anonymous, 1990). In India, maximum use of fungicides is on fruits and vegetables (Table 4.6) followed by the cereals (rice and wheat).

Modern agriculture makes use of a number of toxic and hazardous chemicals, a majority of which are used as biocides. If not used judiciously these chemicals can adversely affect human as well as plant health and ultimately have adverse long term effects on the ecosystem. The primary concern related to the use of these toxic materials is the adverse effect they may exert upon the health as a result of consuming contaminated food or water and upon the long term productivity of the soils. The degree of hazards that a compound may pose is related to its toxicity, persistence or accumulation in soil, plants/plant parts. For many developing countries the present period is critical because if the agricultural use of the pesticides is allowed to develop haphazardly many of the problems that have followed their use in developed countries may be repeated or even exceeded (Sokhi *et al.*, 1993).

Environmental hazards due to the use of pesticides have been studied in some developed countries of Europe and in U.S.A. and as a result of these investigations, there have been restrictions on the use of many of the more persistent pesticides. However, many of these chemicals are still being manufactured in very large quantities in developed countries for export to the developing countries including India.

Herbicides

The use of chemicals for controlling weeds started with the introduction of 2,4-D in 1940's. The usage of herbicides is higher than any other agrochemicals. The main use of herbicides in the world is in maize and soybean.

Farmers have now realised the importance of herbicide usage for harvesting higher crop yields. The use of herbicides during eighties had registered a phenomenal growth as compared to the other pesticides. From a mere 2 per cent of the total pesticide consumption in the early seventies in India, herbicides now account for about 13 per cent of the pesticide market. Herbicide consumption, however, in India is still much less than that in the developed countries, where herbicides constitute 45 per cent of the total pesticide consumption (Sachan, 1989). In to-day's ecologically conscious society, there is strong interest on all aspects of herbicidal use. In fact, the use of herbicides may well depend on educating the public about the ultimate fate of herbicides in the soil, air and water.

The role of soil ecosystem remains relatively poorly understood, but the importance of the soil microflora and fauna lies in the breaking down of organic matter, incorporating it into soil, and releasing nutrients for plant growth. One of the main problems is the complexity of soil ecosystem and the diversity of the soil fauna and flora. Until we understand these better, we have little chance of understanding how chemicals such as herbicides can affect them. The herbicide can have direct effects upon decomposer microorganisms, rhizosphere microorganisms, root pathogens, and disease antagonists. It can also have similar effects on decomposer soil-inhabiting invertebrates, root pests, and pest antagonists, such as parasites and predators, as well as organisms pathogenic to invertebrates. The interaction effects between soil organisms to assess such effects are not known (Edwards *et al.*, 1988). However, soil organisms are such important contributors to key dynamic soil processes such as organic matter breakdown, soil respiration, and nutrient cycling that they can be used as useful indices of the effects of herbicides on soil change inputs of organic matter, in the form of dead weeds, into soil ecosystem. This in turn may accelerate soil erosion.

Herbicides are designed specifically to minimize plant diversity by controlling weeds thus promoting monoculture, so they can also indirectly decrease populations and diversity of related soil organisms and lessen the natural input of organic matter into soil, as well as have direct effects on soil organisms. This may in turn increase the need for inorganic fertilizers and pesticides and magnify the impact of agrochemicals on soil ecosystem.

BIOPESTICIDES FOR INSECT PEST AND DISEASE MANAGEMENT

Biopesticides are natural plant products that belong to the so-called secondary metabolites, which include thousands of alkaloids, terpenoids, phenolics and minor secondary chemicals. These substances have usually no known function in photosynthesis, growth or other basic aspects of plant physiology; however, their biological activity against insects, nematodes, fungi and other organisms is well documented. Every plant species has developed an inbuilt unique chemical complex structure that protects it from pests. The plant kingdom offers us a diverse array of complex chemical structures and almost every imaginable biological activity.

Insect Pest Management

Heavy reliance on pesticides has not worked well in terms of integrated pest management (IPM) of insect pests. Most of the synthetic organic insecticides having high toxicity to pest species are often even more toxic to the beneficial insects. Pesticide residues above tolerance limits have been reported in food and milk and even in human milk (Dhaliwal and Singh, 1993). In view of these facts, the selection of insecticides for use in IPM programmes must take into account their toxicity to animals, human beings and useful organism, their persistence in air, water, soil and the overall environmental impact of their use on agricultural crops (Dhaliwal and Arora, 1994). It is in this context that biopesticides are being considered as environmentally safe, selective, biodegradable, economical and renewable alternatives for use in IPM programmes, especially under organic farming systems.

Neem (*Azadirachta indica*) has emerged as the single most important source of insecticides. According to Saxena (1993) about 100 protolimonoids, limonoids or tetranortriterpenoids, pentanortriterpenoids, hexanortriterpenoids and some nonterpenoids have been isolated from various parts of the neem tree and still more are being isolated. These compounds and neem extracts exhibit almost every type of biological activity conceivable against a wide range of insects. More than 300 species of insects have been reported to be affected by the neem components, (Singh, 1993a). A number of other plant species possessing significant biological activity against insects have been reported during the last decade (Arnason *et al.*, 1989; Ketkar and Ketkar, 1993; Srimannarayana, 1993).

Botanical insecticides are ecologically and environmentally safer and generally affect the behaviour and physiology of insects rather than killing them. Among the botanical pesticides investigated, neem has justifiably received the maximum attention during the last two decades. All parts of the neem tree possess insecticidal property but seed kernel is the most active. Neem bark, leaf, fruit and oil as well as extracts with various solvents especially ethanol have been found to exhibit activity against insect pests. Neem products appear to be quite promising against such nefarious pests as *S. gregaria*, *L. migratoria*, *H. armigera*, *S. litura* and *P. xylostella*. Most of the stored grain pests have also been found to be responsive to neem products.

Plants are rich source of bioactive organic chemicals. It is estimated that the plants may contain as many as 400,000 secondary metabolites. As many as 2121 plant species have been reported to possess pest control properties (Ahmed *et al.*, 1984); 25 of these plant species possess the characteristics required in an ideal botanical insecticide (Ketkar and Ketkar, 1993) and are, therefore, most promising for use in IPM programmes. The plants which have shown potential for use in India in pest management include chinaberry, *Melia azedarach* L.; custard apple, *Annona squamosa* L. and *A. reticulata*; sweetflag, *Acornus calamus*; yellow sage, *Lantana camera* L.; puna oil tree, *Pongamia glabra* Vent., etc. With increasing environmental awareness, pyrethrum has also been considered to play an important role in IPM especially on fruits and vegetable crops. Only a limited number of studies have been conducted on the effect of botanical pesticides on non-target organisms. Saxena (1993) suggests that due to relatively weak contact effect in insects and the unique mode of action, neem-based insecticides have not been found to be detrimental to important natural enemies of insect pests.

Nematode Management

Nematodes normally attack almost all kinds of plants and are important pests of field, fruit and horticultural crops as well as forest trees. Quite often they form disease complexes with other plant pathogens like viruses, bacteria, and fungi. The loss in yield may range from 10 to 40 per cent or even more depending on the susceptibility of host plants, density of nematode population, nutrient availability, soil moisture, temperature and texture. With the introduction of intensive cultivation practices, there has been a growing need for effective crop protection. In view of the nematode problems reported in various crops, synthetic pesticides (nematicides) have been made available and presently used by the farmers.

The nematicides, presently in use are DBCP (1,2- dibromo-3-chloropropane), EDB (Ethylene dibromide), DD (1,3-dichloropropane-1, 2-dichloropropane), nemagon, aldicarb, aladicarb sulfone, phorate, furadan, fensulfothion, nemacur, etc. These nematicides, though effective in reducing nematode populations, are often phytotoxic (Baujard *et al.*, 1989). Nematicides like methyl bromide, dazomet and aldicarb, when applied repeatedly for years, lost their efficacy and could not reduce the population of the pathogens in soil significantly (Hide *et al.*, 1984). Aldicarb residues have been reported to persist in soil for as long as 120 days (Meher *et al.*, 1986). As in the case of other pesticides, nematicides also result in environmental pollution which is ultimately endangering life of soil fauna and fish (Wheler *et al.*, 1979; Landan and Tucker, 1984). Soil fumigation with 1,3-dichloropropane could contaminate ground water (Loria *et al.*, 1986). Residues of aldicarb were reported in coconuts (Singh *et al.*, 1983). Biopesticides, obtained from plants, have shown a great promise in the sense that they are very often non-phytotoxic, easily biodegradable and do not leave toxic residues in the edible products as observed with the use of chemical nematodes.

Oilcakes such as those obtained from mustard (*Brassica canipestrei* L.), Sesame (*Sesamum indicum* L.), Castor (*Ricinus communis* L.) etc. are known to have nematicidal properties. Oil cakes from Undi (*Calophyllum inophyllum* L.) and Sal (*Shorea robusta* Gaerin) were effective in reducing root gall formation and hatching of *Meloidogyne incognita* eggs on tomato seedlings than the oil seed cake of *mahua* (Goswami and Vijay Lakshmi, 1986b). Goswami *et al.* (1988) reported relative efficacy of oil cakes from *Karanj* (*Pongamia pinnata* Pierre), castor (*Ricinus communis* L.), (*Melia azedarch* L.) and carbofuran in reducing root-knot nematode infestation in cowpeas.

Besides oil cakes, other plant materials have proved effective as soil amendments for reducing nematode attack. However, the effective dosage of these materials is quite high as compared to the chemical nematicides. *Meloidogyne incognita* population were reduced due to soil amendment with leaves of *Azadirachta indica*, *Annona squamosa*, *Chrysanthemum coronarium* etc. (Sharma and Tiagi, 1989; Govindaih *et al.*, 1989; Siddiqui and Alam, 1990; Hameed, 1970). Nematicidal principles have also been isolated from different plant species.

Among the nematicides of plant origin the most promising ones are terpenes (mono-and sesqueter penes) and lactones (Chatterjee *et al.*, 1982; Bala, 1987; Sivihabab and Sukul, 1990; Wright *et al.*, 1983 and Milman, 1990).

Disease Management

Neem has also exhibited antifungal activity. Neem oil protected the seeds of chickpea against serious fungal diseases like Rhizoctonia solani, Sclerotium rolfcii and Sclerotinia scleotiorum (NRC, 1992). These fungi which grow widely on various foods are one of the most deadly organisms on earth and the aflatoxins produced by them are highly carcinogenic.

The neem tree addresses to several environmental problems such as the rehabilitation of degraded ecosystem and wastelands, reduction in the use of agrochemicals such as fertilizers and synthetic pesticides and generation of income for small farmers with limited resources. The neem based pesticides are relatively safe and do not leave any residues on agricultural produce. Neem oil has found a wide range of industrial use like that in preparation of soaps, toothpastes, cosmetics etc. Application of neem extracts alongwith urea resulted in a decline in the population of nitrifying bacteria (Prasad *et al.*, 1993).

Biopesticides and other preparations of plant origin used in agriculture seem to have a good scope especially in view of the environmental problems being faced with the synthetic agrochemicals. Their use need to be promoted under the Integrated Pest Management Programmes.

ADVERSE EFFECT ON FOOD QUALITY

Soil-plant transfer of contaminants is the major pathway of contaminant intake by animal and human beings. As discussed above, following entry into the soil environment, many contaminants accumulate in the surface horizon that roots proliferate. Consequently, many of these contaminants are available for plant uptake. Crops such as potatoes and wheat have been shown to accumulate significant amounts of Cd and the concentrations appear to vary with the nature of soil type. This variation in Cd uptake by crops has been to some extent attributed to the presence of chloride which is able to form stable soluble complex with Cd thereby making it more plant available.

$$\text{Soil-Cd]}^{\circ} + nCl^{-} \rightarrow CdCl_n^{2-n} + \text{Soil]}^{2-} \quad \text{....................(12)}$$

Similar to other toxic heavy metals, following animal and human consumption of contaminated food, Cd accumulates in offals. It can interfere with a wide range of biochemical reactions. In recent years regular monitoring of food products for heavy metal accumulation has been carried out in many countries. Table 4.7 gives the information regarding maximum residue limit of food commodities in different countries.

Pesticide contamination of agricultural and horticultural crops have also been reported. Perhaps the most publicised impact of pesticide on food quality is the contamination of Australian beef products with pesticides which was rejected by US Department of Agriculture during 1995. The contaminants used on cotton plants were subsequently used as beef cattle feed. Other examples include DDT presence in human milk (India) and the recent arsenic poisoning of millions of Indians through consumption of As contaminated rice products and drinking of As contaminated water. Arsenic which has chemistry similar to that of phosphorus interferes with biochemical reactions that involve phosphate. On the other hand DDT accumulates in animal fat tissue and gradually with time converts to the more toxic DDE in the presence of enzymes. Such organic substances have been linked to breast cancer. The other pathway of pesticide entry into food chain can be due to inappropriate application rates or insufficient withholding periods prior to crop harvest and marketing. There are few instances where pesticide residues in food crops are a result of plant uptake of the pesticides from soil. For livestock however, ingestion of soil contaminated with pesticides may be a major route for exposure and contamination of the food chain (Harrison *et al.*, 1970).

Table.4.7: Maximum residue limit of food commodities in different countries

Country	Samples exceeding maximum residue limit (%)	
	Domestic food	Imported food
Australia	1.0	-
Canada	0.4	2.3
Denmark	1.3	2.9
Ireland	0.2	0.4
Japan	0.4	-
Philippines	2.8	-
Spain	2.0	-
United Kingdom	1.6	2.1
USA	1.0	3.6

Source: Naidu et al. (1997).

CHAPTER 5

Use of Traditional Additives for Organic Farming

Most countries have traditionally utilized various kinds of organic materials to maintain or improve the tilth, fertility, and productivity of their agricultural soils. However, several decades ago organic recycling practices in some countries were largely replaced with chemical fertilizers which were applied to high yielding cereal crops that responded best to a high level of fertility and adequate moisture, including irrigation. Soil tillage was also intensified to improve weed control and seedbed conditions. Consequently, the importance of organic matter to crop production received less emphasis, and its proper use in soil management was sometimes neglected, or even forgotten. With these changes and the failure to implement effective soil conservation practices, agricultural soils in a number of developed and developing countries have undergone serious degradation and decline in productivity because of excessive soil erosion and nutrient runoff, and decreased soil organic matter levels.

To achieve sufficiency in food production and in maximising crop yield potential, many developing countries will need to accelerate efforts to halt the decline in soil productivity and to restore the productivity of degraded soils in the shortest possible time. Much of this goal can be achieved through proper management and utilization of organic materials such as agricultural residues and municipal wastes to protect agricultural soils from wind and water erosion and to prevent nutrient losses through runoff and leaching. Efficient and effective use of these materials as soil conditioners also provides one of the best means we have for maintaining and restoring soil productivity. The beneficial effects of organic wastes on soil physical properties are widely known as evidenced by increased water infiltration, water-holding capacity, water content, aeration and permeability, soil aggregation and rooting depth, and by decreased soil crusting, bulk density, runoff, and erosion.

SOIL ORGANIC MATTER AND HUMUS

Farmers since ancient times have recognized significant benefits of soil organic matter to crop productivity. These benefits have been the subject of controversy for centuries and some are still debated today. The following list includes many of the recognized benefits of soil organic matter (Stevenson, 1982):

1. It serves as a slow-release source of N, P, and S for plant nutrition and microbial growth.
2. It possesses considerable water-holding capacity, and thereby helps to maintain the water regime of the soil.
3. It acts as a buffer against changes in pH of the soil.
4. Its dark colour contributes to absorption of energy from the sun and heating of the soil.
5. It acts as a "cement" for holding clay and silt particles together, thus contributing to the crumb structure of the soil, and to resistance against soil erosion.
6. It binds micronutrient metal ions in the soil that otherwise might be leached out of surface soils.
7. Organic constituents in the humic substances may act as plant-growth stimulants.

The term soil organic matter is generally used to represent the organic constituents in the soil, excluding undecayed plant and animal tissue, their partial decomposition products, and the soil biomass (Stevenson, 1982). Thus, this term includes-

1. Identifiable high-molecular-weight organic materials such as polysaccharides and proteins,
2. Simpler substances such as sugars, amino acids, and other small molecules, and
3. Humic substances. It is likely that soil organic matter contains

most, if not all, of the organic compounds synthesized by living organisms.

Soil organic matter is frequently said to consist of humic substances and non–humic substances. Non–humic substances are all those materials that can be placed in one of the categories of discrete compounds such as polysaccharides and sugars, proteins and amino acids, fats, simple organic acids, and so on; humic substances are the other, unidentifiable components. Even this apparently simple distinction, however, is not as clearcut as it might appear. Applying this distinction in practice depends on one's ability to separate the humic from the non–humic material and to properly identify the discrete substances. The situation is further complicated by the fact that some identifiable substances are covalently linked to humic material and can be released by hydrolysis or by other means. This situation is another example of an instance where the usage of the terminology is not always clear, but in general, a material can be considered to be a humic substance even though it contains covalently bound fragments of identifiable components.

The term humic substances refer to an operationally defined, heterogenous mixture of naturally occurring organic materials. These substances are ubiquitous in nature and arise from the decay of plant and animal residues in the environment. Humic substances are generally classified into humic acid, fulvic acid, and humin on the basis of their solubility in water as a function of pH. It has long been recognized that humic substances have many beneficial effects on soils and consequently on plant growth. Some of these effects have been well documented, but others, such as the alleged direct effect of humic substances on plant growth, are still a matter of controversy. The inherent complexity of humic substances and the inability of researchers to separate them into pure components impose restrictions on the meaning of the term structure when applied to these materials; we must be satisfied with measuring the average characteristics of these substances. Fortunately, this is sufficient for understanding most of the properties of these materials and for explaining their effects in the soil.

ORGANIC MATERIALS USE

Organic manures, including animal manures, crop residues, green manures, and composts were traditionally and preferentially used in developing countries until the 1960's when inorganic chemical fertilizers began to gain popularity. Chemical fertilizers became easily available and unlike organic manures, they were less bulky and, thus, easier to transport, handle, and store. They were also relatively inexpensive and produced greater crop response than many organic fertilizers. This was particularly true during the Green Revolution when crop varieties were introduced that responded to heavy doses of chemical fertilizers. Thus when the world energy crisis began in early 1970's, chemical fertilizers had virtually replaced organic sources of crop nutrients in some developing countries (FAO, 1975). Because of recent increases in fossil energy costs, chemical fertilizers have become much more expensive, and organic fertilizers have once again started to regain their lost popularity.

A significant consequence of these events, is that "in many countries that have recently been increasingly dependent upon chemical fertilizers, the technical knowledge of organic matter utilization has been lost. It is thus necessary to reintroduce the established techniques, to improve them, and to develop new practices conforming to modern technology."

The shift away from organic recycling practices also served to reemphasize the value of, and need for, organic amendments for short-and long-term improvement of cultivated soils and maintenance of soil productivity. Without regular additions of adequate amounts of organic materials to soils, there is often a concomitant increase in nutrient runoff, erosion, and gradual deterioration of soil physical properties.

Environment pollution has also become an international concern. Proper processing and recycling of organic wastes and residues as resources for agriculture can greatly reduce environmental pollution. Additional benefits include improved public health, conservation of resources, and better appearance of both urban and rural communities.

Reintroduction of Best Management Practices

Many farmers in developing countries have shifted towards intensive cropping for short-term economic gain, often at the expense of soil and water conservation practices. This has resulted in markedly increased runoff and enrichment of surface waters. With increasing public concern for the role of agriculture in environmental pollution, there is an urgent need to utilize those soil ad crop management practices which have been identified as best management practices for controlling soil erosion and water pollution from cropland. These include the use of sod-based rotations, contouring, conservation tillage, cover crops, grassed waterways, and possibly others such as divided-slope farming. Research and extension programmes should be formulated and implemented, in both developed and developing countries, to demonstrate the cost/benefit relationships of conservation management practices, Multiple cropping systems such as double cropping, sequential cropping, and intercropping may also provide special means for controlling wind and water erosion, and for effective recycling of nutrients from crop residues and other organic wastes.

Composting to Enhance the Usefulness and Acceptability of organic Materials

Composting is one way in which some of the problems associated with the utilization of various organic wastes (e.g., odors, human pathogens, and storage and handling constraints) can be resolved. Composting is an ancient practice whereby farmers have converted organic wastes into resources that provide nutrients to crops and enhance soil tilth, fertility, and productivity. Through composting, organic wastes are decomposed, nutrients are made available to plants, pathogens are destroyed, and malodours are abated. The historical aspects of composting have been thoroughly discussed in reviews by Gotaas, 1956; Golueke, 1972.

Recently, the USDA at Beltsville, Md., developed the highly successful Beltsville Acrated Rapid Composting (BARC) Method for composting sewage sludge, animal manures, municipal refuse, and pit latrine wastes (Willson *et al.*, 1980). This method has been widely adopted by both large and small municipalities throughout the USA for composting sewage sludge and solid waste. A number of developing countries have also adopted this technology for composting. The method is simple and relatively inexpensive, yet effective, and allows considerable trade-off between labour and capital.

Composts provide a more stabilized form of organic matter than do raw wastes, and can vastly improve the physical properties of soils. For example, addition of sludge compost to sandy soils will increase their ability to retain water and render them less droughtly. In heavy-textured clay soils, the added organic matter will increase permeability and water infiltration and clay soils have been shown to reduce soil compaction, lower the bulk density, and increase the rooting depth (Mays *et al.*, 1973; Epstein, 1975; Epstein et. al., 1976.

Hornick *et al.*, (1979) have discussed the uses of sewage sludge compost for soil improvement and plant growth including (i) establishment, maintenance, and production of turfgrass and sod, (ii) use in vegetable gardens, (iii) production of field crops and forage grasses, (iv) use on nursery crops and ornamentals, (v) use in potting mixes, and (vi) reclamation and revegetation of disturbed lands. Recommendations are provided as to time, methods, and rates of compost application for different soils and management practices.

Some organic materials may have chemical, physical, and/or microbiological properties that would greatly limit the extent to which they could be composted alone. For example, some wastes have an extremely acidic or alkaline pH, others may have an unusually high or low C : N ratios, and still others may vary widely in their solid content. In such cases, selective co-composting of these wastes with sewage sludges, pit latrine waste or night soil, municipal solid waste (*i.e.*, garbage or refuse), crop residues, animal manures, food processing wastes or certain industrial wastes, may alleviate these deficiencies and provide a readily compostable mixture and higher quality product.

Consideration of Organic Methods of Farming

According to a recent report on organic farming by the U.S. Department of Agriculture (USDA, 1980), there are a number of farmers in the USA operating large-and small-scale commercial farms profitably with minimal or no use of chemical fertilizers or pesticides. They are often referred to as organic farming methods, they collectively advocate that sustainable and successful farming systems are based primarily on the proper care and protection of the soil.

Organic farming employs basically a systems approach to farm management. On well-managed farms the various practices used are often interrelated so that each augments the other to form a complex but efficient production system. For example, a legume may be grown in a crop rotation not only to produce feed for animals but to control specific weeds and insects, and to supply N for grain crops that follow. Most organic farmers in the USA rely heavily on recycling of organic materials and use of green manure crops and legumes inter rotation to supply nutrients, and to maintain nutrient balances and soil organic matter. As a group, organic farmers are highly committed to protecting the soil resource which they do by regular use of sod-based rotations, legumes, animal manures, and other organic wastes. The USDA report (1980) revealed that organic farming is also practiced to a limited extent in other developed countries, including Japan and those of Western Europe.

Special Management Practices for Utilization of Organic Materials on Land

Special management practices may be needed if we are to obtain the full value of organic wastes as soil amendments and as sources of plant nutrients. For example, the value of organic materials as fertilizers is increased if nutrient release through decomposition and mineralization coincides with the crop's nutrient demand curve. The nutrient release pattern of organic materials can be controlled to some extent by proper timing of application, pretreatment methods such as composting, and by the method of application. For example, certain organic materials may decompose more slowly if left on the soil surface or concentrated as bands in the soil than if mixed thoroughly in the tillage layer (Parr and Papendick, 1978).

There are many problems involved in handling and applying organic materials to land because their physical characteristics are so variable. Organic wastes and residues are almost always bulky, and can exist as liquids, slurries, sludges, or solid materials. Available application technologies are often ineffective and fail to achieve the desired level of erosion control or increased crop production. Methods of application must be simple, inexpensive, energy conserving, and effective for nutrient recycling and erosion control.

An example of a particularly effective soil erosion control measure for sloping lands is that of vertical mulching, a procedure which incorporates organic materials such as crop residues into a vertical channel 30 to 40 cm deep and 10 to 12 cm wide at the soil surface (Parr, 1959). Channels are established on the contour with intervals ranging from 5 to 10 m. A recent modification of this procedure is referred to as slot mulch and has been demonstrated as an effective means of controlling erosion on steeply sloping lands in the winter wheat area of eastern Washington State (Saxton *et al.*, 1981). Both techniques are highly effective in intercepting runoff, enhancing root penetration and development, and providing for effective water conservation and storage for crop use. Although several types of machines have been designed for these operations, similar results could be achieved using hand labour.

This concept of residue management may have application in developing countries for restoring the productivity of sloping lands that have suffered from severe soil erosion. Besides crop residues, other types of organic materials, including composts, could also be utilized in the vertical mulch or slot mulch procedures.

Value of Organic Materials as Fertilizers and Soil Conditioners

The value of organic materials as fertilizers and soil conditioners is often misunderstood and has been the source of some controversy. The simplest and most common means of estimating the value of organic amendments is by assessing the current market value of the potentially available plant nutrients they contain. Usually this is done in terms of their macronutrient content, i.e., N, P, and K.

However, many organic materials contain other components which can contribute significantly to increased crop yields, including organic matter, secondary and micronutrients, and sometimes lime. In some cases, the organic matter fraction of a particular material may be of greater value than its total nutrient content because of the beneficial effect of organic matter on soil physical properties and improvement of soil productivity.

Bulky Organic Manures

Farmyard manure (FYM), farm compost, town compost, night soil, sludge and green manures are bulky in nature and supply large quantities of organic matter but small quantities of plant nutrients in comparison to the inorganic fertilizers.

Farmyard Manure

Good-quality farmyard manure is perhaps the most valuable organic matter applied to a soil. It is the most commonly used organic manure in most countries of the world. It consists of a decomposed mixture of cattle dung, the bedding used in the stable and any ramnants of straw and plant stalks fed to cattle. It is one of the most important agricultural by-products. Unfortunately more than 50 per cent of the cattle dung produced in India is burnt as fuel and is thus lost to agriculture.

TABLE 5.1: Average nutrient content of bulky manures

Manure	Percentage content		
	Nitrogen (N)	**Phosphoric acid (P_2O_5)**	**Potash (K_2O)**
Animal refuse			
Cattle dung, fresh	0.3-0.4	0.1-0.2	0.1-0.3
Horse dung, fresh	0.4-0.5	0.3-0.4	0.3-0.4
Sheep dung, fresh	0.5-0.7	0.4-0.6	0.3-1.0
Night–soil, fresh	1.0-1.6	0.8-1.2	0.2-0.6
Poultry manure, fresh	1.0-1.8	1.4-1.8	0.8-0.9
Raw sewage, fresh	2.0-3.0	-	-
Sewage sludge, dry	2.0-3.5	1.0-5.0	0.2-0.5

Manure	Percentage content		
	Nitrogen (N)	Phosphoric acid (P_2O_5)	Potash (K_2O)
Sewage sludge, activated dry	4.0-7.0	2.1-4.2	0.5-0.7
Cattle urine	0.9-1.2	tr.	0.5-1.0
Horse urine	1.2-1.5	tr.	1.3-1.5
Human urine	0.6-1.0	0.1-0.2	0.2-0.3
Sheep urine	1.5-1.7	tr.	1.8-2.0
Wood ashes			
Ash, coal	0.73	0.45	0.53
Ash, household	0.5-1.9	1.6-4.2	2.3-12.0
Ash, wood	0.1-0.2	0.8-5.9	1.5-36.0
Farm, factory and habitation wastes			
Rural compost, dry	0.5-1.0	0.4-0.8	0.8-1.2
Urban compost, dry	0.7-2.0	0.9-3.0	1.0-2.0
Farmyard manure, dry	0.4-1.5	0.3-0.9	0.3-1.9
Filter-press cake	1.0-1.5	4.0-5.0	2.0-7.0
Plant residues			
Rice hulls	0.3-0.5	0.2-0.5	0.3-0.5
Groundnut husks	1.6-1.8	0.3-0.5	1.1-1.7
Straw and stalks			
Pearlmillet	0.65	0.75	2.50
Banana, dry	0.61	0.12	1.00
Cotton	0.44	0.10	0.66
Sorghum	0.40	0.23	2.17
Maize	0.42	1.57	1.65
Paddy	0.36	0.08	0.71
Tobacco	1.12	0.84	0.80
Pigeonpea	1.10	0.58	1.28
Wheat	0.53	0.10	1.10
Sugarcane trash	0.35	0.10	0.60
Tobacco dust	1.10	0.31	0.93
Tree leaves, dry			
Calotropis gigantea	0.35	0.12	0.36
Careya arborea	1.67	0.40	2.20
Cassia auriculata	0.98	0.12	0.67
Dillenia pentagyna	1.34	0.50	3.20
Madhuca indica	1.66	0.50	2.00

Manure	Percentage content		
	Nitrogen (N)	**Phosphoric acid (P_2O_5)**	**Potash (K_2O)**
Pengamia pinnata	3.69	2.41	2.42
Pterocarpus marsupium	1.97	0.40	2.90
Terimalia chebula	1.46	0.35	1.35
Terminalia paniculata	1.70	0.40	1.60
Terminalia tomentosa	1.39	0.40	1.80
Xylia dolabriformis	1.37	0.30	1.61
Green manures, fresh			
Cowpea (*Vigna catjang*)	0.71	0.15	0.58
Sesbania aculeate	0.62	-	-
Cluster-bean (*Cyamopsis tetragonoloba*)	0.34	-	-
Horse-gram (*Dolichos biflorus*)	0.33	-	-
Mothbean (*Vigna aconitifolia*)	0.80	-	-
Greengram (*Vigna radiata*)	0.72	0.18	0.53
Sunnhemp (*Crotalaria juncea*)	0.75	0.12	0.51
Blackgram (*Vigna mungo*)	0.85	0.18	0.53

Not only this is a tremendous waste, but also the traditional method of preparing and storing the farmyard manure is generally faulty. The cattle-dung, together with stable-waste and house sweepings, is first collected in the open backyard, and when a cartload has been collected, it is removed to another heap or to an uncovered pit in a common plot outside the village. The loose heaps lie exposed to the sun, with the result that the raw organic matter dries up quickly and does not rot properly. Very often, a part of the dry dung is washed away by rain. Cattle urine is either not conserved or is stored in a defective manner.

The wastage of nitrogen-rich urine, the loss of nitrogen (in the form of ammonia) due to the fermentation of exposed cattle dung, and the washing away of soluble mineral elements by leaching reduce the manurial value of FYM to a great extent. The average composition of plant nutrients in FYM under Indian and European conditions is given below:

	Nutrient Composition (%)		
	N	P_2O_5	K_2O
India	0.3	0.15	0.3
European countries	1.0	0.30	1.0

About half of the nitrogen, one-sixth of phosphorus and more than half of potash are readily soluble and subject to dissipation. The loss of nitrogen and other manures caused by careless handling can be reduced greatly by using an absorbent bedding for cattle, storing dung in stone or brick-lined pits, mixing large quantities of straw and other vegetable matter with cattle dung, and keeping the heap compact and moist.

If urine is not conserved in the bedding used for cattle it must be collected in a covered pucca cistern and then, added to the dung in the manure pit. Nitrogen in the urine is mainly in the form of urea, which readily changes into the highly volatile- ammonium carbonate through bacterial action, and quickly loses ammonia thereafter by evaporation. This loss can be reduced to a great deal if the manure and the urine-soaked absorptive litter for bedding are kept compacted in a pit. The pit may be 1 m in depth, 1.3 to 1.5 m in width and 4.5 to 6 m in length, depending upon the number of cattle on a farm. The filling of the pit should be sectional and when each section of 1.3 m in length is filled to about 45 cm above the ground level, it should be plastered with 2.5 cm layer of a mixture of mud and dung in equal proportions. Before plastering, four to five buckets of water should be added to the manure in the pit. Plastering conserves moisture and nitrogen and also prevents the housefly nuisance. The manure becomes ready for use in about four to five months after plastering.

The quality of manure is also improved by the concentrated feeds given to cattle. Cotton-seed, cotton-seed cake, linseed-meal, wheat bran, grain husk, groundnut cake, gram, horse-gram, etc. are rich in nitrogen, phosphorus, potassium, magnesium and sulphur. It has been found that in the case of adult working-cattle, about 80 per cent of nitrogen and the other mineral elements contained in the feed is recovered in urine, faeces and other animal by-products. Accordingly, manure from cattle fed on cereal straws and grass hay is much less valuable than that from animals fed on legume hays, grains and concentrates.

Considerable attention has also been given to the use of preservatives to manure. Calcium sulphate or gypsum and superphosphate have proved most promising in preventing the escape of ammonia. Gypsum has been found specially effective as an ammonia-absorbing agent. Superphosphate, besides absorbing ammonia, supplies additional phosphorus and thus, improves the crop-producing capacity of the manure.

Partially rotten farmyard manure should generally be applied to the soil about three to four weeks before the sowing of a crop. In case there is sufficient moisture in the soil, there will be enough time for its decomposition and for improving the soil structure. Its application too long before sowing a crop will either cause a drying up of the rotted manure or too quick decomposition, depending on the incidence of rains. But in each case, there will be a serious loss of ammonia and nitrogen. If the manure is already well rotted, it is advisable to apply it just before sowing to a crop. This procedure is particularly essential in the case of light soils. In any case, after the manure is carted to the field, it should be evenly spread and worked into the soil soon to avoid the loss of nitrogen. The practice of leaving the manure in small heaps scattered in the field for several days, before spreading it in the field and incorporating it into the soil, results in the serious deteriortion of its quality, particularly if strong winds blow. In vegetable and fruit cultivation, the application of well-rotten manure, in conjunction with fertilizers to young plants individually, has been found to give the best results.

The practice of penning cattle, sheep and goats in the fields in summer is common in some parts of India. Folding 7,000 sheep for one night is said to add the equivalent of 149.3 quintals (14.93 tonnes) of cattle dung. The fresh dung left in the field in such cases rapidly dries up. This drying checks ammonification and loss of nitrogen. With the first fall of rain, the dung is worked into the soil. It, therefore does not lose much of its fertilizing value. Further more, its beneficial effects on the physical condition of the soil are undeniable. However, sheep-folding is said to make the land more weedy.

There must be adequate moisture in the soil for the proper decomposition of organic matter. Farmyard manure can, therefore be applied to all crops grown in the rainy season or grown under irrigation. The quantity of manure to be applied to un–irrigated crops varies from 1½ to 2 cartloads per hectare in areas of heavier rainfall. If sufficient farmyard manure is not available, it may be applied at the usual rate to a part of the land, say, to one-third or one-fourth of the area, in rotation every year, so that all parts of the field receive the manure regularly once in three or four years. For irrigated field crops, the rate varies from 10 to 20 cartloads. Sugarcane, maize and garden crops, such as potatoes, turmeric, ginger, vegetables and fruits receive still higher doses, amounting sometimes to 15 to 25 cartloads. A cartload of manure, measuring 9 cubic metres, weighs about half a tonne.

It must be stressed that the value of farmyard manure in soil improvement is due to its content of macronutrients and its ability to (i) improve the soil tilth and aeration, (ii) increase the water-holding capacity of the soil, and (iii) stimulate the activity of micro-organisms that make the plant-food elements in the soil readily available to crops. The supply of organic matter, which is later converted into humus is a property of farmyard manure. One tonne of cattle dung can supply only 2.95 kg of nitrogen, 1.59 kg of phosphoric acid and 2.95 kg of potash.

The use of farmyard manure alone causes an imbalance in nutrition owing to its relatively low content of phosphoric, essential elements of plant food in a readily available form, and also to keep the soils well supplied with all the essential elements of plant food in a readily available form. It is beneficial to use the bulky organic manures in conjunction with superphosphate and such other fertilizers as contain the particular plant nutrients in which a soil may be deficient or which the crop to be grown may specially require.

Composted Manure

Another method of augmenting the supplies of organic matter is the preparation of compost from farmhouse, and cattle-shed wastes of all types. Composting has been advocated and adopted extensively during the past 25 years. Composting is the process of reducing vegetable and animal refuse (rural or urban) to a quickly utilizable condition for improving and maintaining soil fertility. Studies conducted in India and elsewhere have shown that good organic manure similar in appearance and fertilizing value to cattle manure can be produced from waste materials of various kinds, such as cereal straws, crop stubbles, cotton stalks, groundnut husk, farm weeds and grasses, leaves, leaf-mould, house refuse, wood ash, litter, urine-soaked earth from cattle-sheds and other similar substances. These materials are rich in cellulose and other readily decomposable carbohydrates and have a carbon-nitrogen ratio of 40 or more to 1. The direct application of such undecomposed, low-nitrogen organic matter as manure brings about a temporary deficiency of mineral nutrients (specially nitrates and ammonium compounds) in the soil by stimulating the growth of micro-organisms, which in turn, compete with crop plants for available nitrogen, phosphorus and other elements. Hence, before using them as manure, it is necessary to compost or partially decompose them. This process lowers the carbon-nitrogen ratio to about 10 or 12 to 1.

Two methods of composting waste organic materials are usually recommended. One depends on aerobic and the other on anaerobic decomposition. In both cases, the farm wastes have to be used as a bedding for cattle in order to absorb a large portion of the animal urine. In the aerobic process, the used bedding, the sweepings from cattle-sheds and some urine-soaked earth from the stable floor are removed every day, mixed with a little cattle dung and two or three handfuls of wood ash are deposited on a well-drained site to gradually build up a low pile, about 30 to 45 cm in height, about 5 m in width and of any convenient length. The pile is built up before the start of the rainy season. After the first heavy showers, the wetted material in a 1.2 m strip on each side of the long heap is turned with a rake on to a 2.4 m wide strip in the middle, thus raising the height of the heap to nearly one metre. This process prevents a loss of moisture and ensures a quick start of decomposition. When the heap sinks appreciably and such a sinking takes about three to four weeks, it is given a turning and made into a fresh heap, thus mixing the outside material with that from inside. After about a month or more, depending on the incidence of rains the heap is given a final turning on a cloudy or moderately rainy day and rebuilt in the vacant part of the original position. The compost becomes ready for use in about four months. This method is quite suitable for composting in the rainy season. The following proportion of raw materials is considered suitable for compost-making-

Raw Materials	**Parts by weight**
Mixed farm residues and cattle-shed wastes	400
Urine-soaked earth	56
Fresh cattle dung	60
Wood ashes	6

If urine soaked earth and cattle-dung are not available, the raw organic materials can still be composted, provided more than one-third of the residues are soft and finely broken, such as fallen leaves, leaf-mould, kitchen wastes, grass clippings, green and succulent weeds, plant trimmings and chopped wheat, barley and soft cereal straw. The use of ordinary soil and wood ash or lime is, however, essential.

In the anaerobic process, the mixed farm residues are collected in pits of a convenient size, say, 4.5 m x 1.5 m x 1 m. Each day's collection is spread in a thin layer, sprinkled with a mixture of fresh cowdung (4.50 kg), ash (140 to 170 g) and water (18 to 22 litres) and compacted. The pit is filled till the raw material stands 38 to 46 cm above its edge, and is then plastered with a 2.5 cm layer of a mixture of mud and cowdung. Under such conditions, decomposition is anaerobic and high temperatures do not develop. Insoluble nitrogen compounds gradually become soluble and the carbonaceous matter is broken down into carbon dioxide and water. The loss of ammonia is negligible, because in high concentration of carbon dioxide. The plastered pit also prevents the fly nuisance. The compacted, moist material becomes composted in about four to five months without any further attention. The well-made compost contains 0.8 to 1 per cent of nitrogen and has all the good properties of farmyard manure. It can be used in the same way as the later. The anaerobic process is particularly suited for use by gardeners in or near cities and towns.

Organic manures from animal wastes play an important role in soil fertility in association with chemical fertilizers. Due to expensive chemical fertilizers, animal wastes are regaining importance as organic fertilizers. According to an estimate, the production of dung and urine from bovine population in India works out to be 1002 and 658 million tonnes, respectively contributing about 5.71 million tonnes of NPK based on the average percentage of nutrient in dung and urine (Gaur *et al.*, 1984).

Urban Compost

In recent years, large-scale composting of town refuse and night-soil in properly constructed trenches away from human habitations has been taken up successfully by the municipal bodies of many large and small towns. Trenches, 1 to 1.2 m wide, 75 cm deep and of convenient length, are filled with successive layers of night-soil, town refuse and earth, in this order. The compost gets ready in about three months. The following figures of volume-weight conversion will be found useful in the preparation of town compost.

Volume	Weight in kg or qtls
1 cu m of refuse	318 kg or 3.18 qtls
1 litre of night soil	0.991 kg
1 cu m of compost	636 kg of 6.36 qtls
1 cartload of refuse (0.849 cu m)	95.40 kg

Social prejudice against the use of this valuable compost has disappeared and town-composting is almost being rapidly adopted in most of the countries.

Sewage and Sludge

Domestic and industrial wastes (sewage and sludge) contain large quantities of plant nutrients and are used for growing of crops near many towns. In many places, the undiluted sludges has been found to be too strong for healthy plant growth and if it contains readily oxidizable organic matter, its use actually reduces nitrates present in the soil. The disadvantages are still greater if sewage is used on land without preliminary treatment. The soil quickly becomes sewage sick owing to the mechanical clogging by colloidal matter in the sewage and the development of anaerobic organisms which not only reduce the nitrates already present in the soil but also produce alkalinity. Bacterial contamination makes the eating of raw vegetables grown on untreated sewage or sewage sludge a real danger to health. On dry-weight basis, contains 3 to 6 per cent N, about 2 per cent P2O5 and 1 per cent K2O in forms that can become readily available when applied to the soil. Similarly, the effluent is a clear, odourless liquid containing nitrates in solution, from which most of the pathogenic bacteria originally present have been removed. Both the activated sludge and the effluent can be used with safety for manuring and irrigating crops. However, under no circumstances should any produce grown on a sewage-farm be eaten uncooked.

Municipal and sewage wastes also form an important component of organic farming. The total and utilisable nutrient potential from garbage and sewage sludge works out to be substantial. The total municipal refuse is about 12 million tonnes/annum containing about 0.5% N, 0.3% K, whereas sewage sludge amounting to 4 million tonnes/annum contains about 3% N, 2% P and 0.3% K (Bhardwaj and Gaur, 1985). Besides the necessity of sewage farming from nutrient utilisation point of view, it is also an effective method to avoid pollution.

Night-Soil

Very few towns in India are equipped with proper sewage system. The sanitary disposal of night-soil with an effective control of foul smell and fly nuisance is, therefore, a serious problem all over the country. Since human excreta is a potential source of soil improvement, public health authorities in several countries make the necessary arrangements for its conservation and conversion into a form in which it can be safely used as a manure. The dehydration of night-soil, as such, or after admixture with absorbing materials, e.g. soil, ash, charcoal, and sawdust, produces a poudrette that can be easily used as a manure. The mixing of night-soil with an equal volume of ash and 10 per cent powdered charcoal produces an odourless material, containing 1.32 per cent nitrogen, 2.8 per cent phosphoric acid, 4.1 per cent potash and 24.2 per cent lime.

The addition of 40 to 50 per cent of sawdust to the night-soil yields straightway a dry, acidic poudrette which may contain 2 or 3 per cent nitrogen.

Green Manuring

Green manuring can be defined as a practice of ploughing or turning into the soil undecomposed green plant tissues for the purpose of improving physical structure as well as fertility of the soil. From the time immemorial the turning in a green crop for improvement of the conditions of the soil has been a popular farming practice.

Despite special efforts at increasing the supplies of farmyard manure and compost, the supply of farm and other organic manures is scarce and ever becoming more costly. Green-manuring, wherever feasible, is the principal supplementary means of adding organic matter to the soil. It consists in the growing of a quick-growing crop and ploughing it under to incorporate it into the soil. The green-manure crop supplies organic matter as well as additional nitrogen, particularly if it is a legume crop, which has the ability to fix nitrogen from the air with the help of its root-nodule bacteria. A leguminous crop producing 8 to 25 tonnes of green matter per hectare will add about 60 to 90 kg of nitrogen when ploughed under. This amount would equal an application of 3 to 10 tonnes of farmyard manure on the basis of organic matter and its nitrogen contribution. The green-manure crops also exercise a protective action against erosion and leaching.

The crops most commonly used for green-manuring are the following:

Sunnhemp (*Crotalaria juncea*), dhaincha (*Sesbania aculeata*), cluster-bean (*Cyamopsis tetragonoloba*), senji (*Melilotus parviflora*), cowpea (*Vigna catjang*, *Vigna sinensis*), horse-gram (*Dolichos biflorus*), pillipesara (*Phaseolus trilobus*), berseem or Egyptian clover (*Trifolium alexandrinum*).

Types of Green Manuring

The adoption of green manuring depends upon the agroclimatic conditions. Broadly the following two types of green manuring can be thought of.

Green manuring in situ: In this system, green manure crops are grown and buried in the same field which is to be green-manured, either as a pure crop or as an intercrop with the main crop. The most common green manure crops grown under this system are sunnhemp (*Crotalaria juncea*), dhaincha (*Sesbania aculeata*), pillipesara (*Phaseolus trilobus*) and clusterbean (*Cyamopsis tetragonoloba*).

In India, various methods of growing green manure crops in situ are followed to suit local conditions. For the proper decomposition of the green manure, it is necessary that the green material should be succulent and there should be adequate moisture in the soil. Plants at the flowering stage, contain the greatest bulk of succulent organic matter with a low carbon/nitrogen ratio. The incorporation of the green-manure crop into the soil at this stage allows a quick liberation of nitrogen in the available form. With advancing age, the percentage of carbonaceous matter in the plants increases and that of nitrogen decreases. If the material with a wide carbon/nitrogen ratio is ploughed under, micro-organisms bring about its decomposition, draw up the released nitrogen and other nutrients and cause a temporary nutrient deficiency.

The increase of yield after green-manuring is usually of the order of 30 to 50 per cent. The fertilizing value of the legume crop can be increased a great deal by manuring it with superphosphate application. This practice not only increases the phosphorus content of the green-manure plants, but also encourages their plant growth on the whole, thus converting an inorganic fertilizer into an organic manure. Green manures also have an appreciable residual effect.

Green leaf manuring: Green-leaf manuring refers to turning into the soil green leaves and tender green twigs collected from shrubs and trees grown on bunds, waste lands and nearby forest areas. The common shrubs and trees used are Glyricidia (*Glyricidia maculata*), *Sesbania speciosa*, Karanj (*Pongamia pinnata*), etc.

TABLE 5.2: Characteristics of some green manuring crops

Crops	Green matter (t/ha)	Moisture (%)	N content (% dry wt)
Cultivated annual legumes			
Cassia mimosoides	4.7	74	2.99
Crotalaria juncea	16.5	73	3.01
Cyamposis tetragonoloba	7.0	60	3.53
Indigofera anil	6.8	74	3.66
Sesbania aculeata	14.8	78	2.67
Sesbania speciosa	7.8	78	2.43
Vigna unguiculata	10.0	85	2.63
Vigna radiata var aureus	7.7	82	2.96
Vigna trilobus	5.3	79	2.47
Perennial legumes			
Cassia hirsuta	5.0	81	2.52
Desmodium ovroides	1.4	72	3.35
Gliricidia maculata	3.0	75	3.46
Sesbania punctata	3.7	73	2.42
Tephrosia candida	2.3	67	3.20
Wild annual legumes			
Aeshynomene americana	8.9	78	3.14
Calopogonium mucunoides	4.5	74	3.02
Cassia tera	5.2	71	2.13
Cassia occidentalis	4.3	78	2.80
Lathyrus sativus	-	82	4.67
Tephrosia purpurea	3.5	70	3.46

Green manuring *in situ* is followed in northern India while green leaf manuring is common in eastern and central India. It is generally practised for supply of nitrogen to crop plants, although it also provides organic matter per se and other plant nutrients in soil. The best known and popular green manures are those which fix atmospheric nitrogen in association with Rhizobium. A legume with greater leafy growth, the ability to suppress weeds and a well developed root system may be the best form of green manure. A large number of green manure crops have been listed by Panse *et al.* (1965). Of the annual cultivated legumes, *Crotalaria juncea* and *Sesbania cannabina* (syn: *S. aculeata*) have been most widely grown by farmers in Asia because of their better adaptability under flooded condition (Gaur *et al.*, 1984; Ladha *et al.*, 1988).

Investigations conducted in sandy loam soil of Delhi showed a beneficial effect of green manuring with and without phosphorus on wheat crop. Total N, available N and P, and bacterial and *Azotobacter* populations in the soil, were increased significantly. The percentage of water stable aggregates was also enhanced. The combined effect of green manuring and phosphorus fertilization was superior to green manuring alone (Kute and Mann, 1968).

Singh (1967) reported that the plots fertilized with inorganic sources of NPK gave only slightly higher yields than green-manured plots with an annual dressing of phosphorus. Green manuring was not of much use in increasing organic matter content of the soil to higher level but contributed towards the supply of nitrogen and other plant nutrients to the crop.

Sesbania rostrata, a tropical legume colonising water-logged soils in Sahel region of West Africa, forms nitrogen fixing nodules with Rhizobium sp. on root as well as on stem (Dreyfus and Dommergues, 1981). It is one of the three genera of leguminoseae which bears stem nodules in addition to root nodules. Other two being *Aeschynomene* belonging to family Papilionoideae and *Neptunia* of Mimosoideae sub-family. Unlike *Aeschynomene indica* and *Neptunia obraceae*, *Sesbania rostrata* shows profused stem nodulation. In field, the plants attain height upto 4 metres. Stem nodules are formed along the lines of stem lenticals upto 2.5 m above ground level. Due to presence of chloroplast in the peripheral cortex, the nodules are green. Recently, *Sesbania rostrata* has gained much significance because of its extremely fast growing nature and nitrogen fixing abilities under flooded condition.

The comparative effects of *Sesbania rostrata* and *Sesbania cannabina* as a green manure crop for rice in alluvial soil was investigated in greenhouse conditions (Gaur and Singh, 1989). The study showed that *Sesbania rostrata* nodulated both in stem and roots when incorporated @ 10 t/ha gave the maximum growth of rice plants showing increased height, tillers, root and aerial biomass and crop yield. When only root nodulated *Sesbania cannabina* or *Sesbania rostrata* were incorporated in the soil, their effect was significantly less than *Sesbania rostrata* bearing both stem and root nodules. Nitrogen uptake by the plant was also higher in presence of *Sesbania rostrata* as compared to *Sesbania cannabina*.

Green-Leaf Manuring

Green leaves from trees, viz. Thespesia populnea, Cassia auriculata, Pongamia glabra, Melia azadirachta, Calotropis gigantea, Adhatoda vasica, etc. are collected and used for green manuring in sourthern parts of India. Weeds, such as Croton sparsifolorus, Lucas aspera and Stachytarpheta indica are also utilised for green-leaf manuring. Kumaraperumal et al. (1975) reported that the incorporation of green leaf (Glyricidia maculata and Pongarnia glabra) manure on the yield of sugarcane seems to be better and comparable with complete manuring with inorganic fertilizers. The juice quality was not affected by the form of manuring.

Some wild legume plants, viz. *Gliricidia maculata*, *Pongamia glabra*, *Calotropis gigantea*, *Tephrosia purpurea*, *Tephrosia candida*, *Ipomoea carnea*, *Cassia tora*, *Sesbania* sp and *Indigofera teysmann* can be grown on bunds and wastelands for utilising their vegetative parts for green-leaf manuring.

Concentrated Organic Manures

Concentrated organic manures are those materials that are organic in nature and contain higher percentages of essential plant nutrients such as nitrogen, phosphorus and potash, as compared to bulky organic manures. These concentrated manures are made from raw materials of animal or plant origin. The concentrated organic manures commonly used are oilcakes, blood-meal, fish meal, meatmeal and horn and hoof meal.

Oilcakes

Oilcakes can be grouped into two classes, namely. (i) edible oilcakes- suitable for feeding the cattle and (ii) non-edible oilcakes- not suitable for feeding the cattle. Edible oilcakes, are normally fed to cattle as concentrates. However, some of the edible oilcakes are also applied by the farmers to the soil.

An estimated 2.5 million tons of oilcakes are produced annually in India. Oilcakes contain nitrogen, phosphorus and potash, along with a large quantity of organic matter. The chemical composition of the major oilcakes available in the country is given in Table 5.3.

The quantity of nitrogen varies with the type of oilcake. It is from 2.5 per cent in mahua cake to 7.9 per cent in decorticated safflower cake. In addition to nitrogen, all oilcakes contain small percentages of P_2O_5 (0.8 to 2.9 per cent) and potash (1.2 to 2.2 per cent). Oilcakes are the quick-acting organic manures. Though they are insoluble in water, their nitrogen becomes quickly available to the plants in about 7 to 10 days after application. The *mahua* cake, however is an exception, as its nitrogen does not become available till about two months after application. As such *mahua* cake should be applied about two months before sowing, provided the soil is moist. This cake is suitable for application to fruit plants and plantation crops.

TABLE 5.3: Average nutrient content of major oilcakes used in India

Oilcake	Percentage composition		
	N	P_2O_5	K_2O
Non-Edible Oilcakes			
Castor cake	4.3	1.8	1.3
Cottonseed cake undecorticated	3.9	1.8	1.6
Karanj or honge cake	3.9	0.9	1.2
Mahua or ippi cake	2.5	0.8	1.8
Neem cake	5.2	1.0	1.4
Safflower cake undecorticated	4.9	1.4	1.2
Undi cake	3.6	1.5	2.0
Edible Oilcakes			
Coconut cake	3.0	1.9	1.8
Cotton seed cake decorticated	6.4	2.9	2.2
Groundnut cake	7.3	1.5	1.3
Linseed cake	4.9	1.4	1.3
Jambo cake	4.9	1.6	1.9
Niger cake	4.7	1.8	1.3
Rape seed cake	5.2	1.8	1.2
Safflower cake decorticated	7.9	2.2	1.9
Sesame or til cake	6.2	2.0	1.2

Oilcakes should be well-powdered before application, so that they can be spread evenly and are easily decomposed by microorganisms. They can be applied a few days prior to sowing or at sowing time. Oilcakes-especially groundnut cake, are also applied extensively in the form of a top-dressing to sugarcane. Depending on crop, oilcakes are applied broadcast, drilled or placed while earthing up near root zone.

Oilcakes from country oil extraction units contain a little more oil than the hydraulic or expeller-pressed oilcakes. Due to the higher percentage of oil, oilcakes from country units are somewhat slow-acting, as oil prevents rapid conversion of organic nitrogen into available form.

Meatmeal

An adult animal can provide 35 to 45 kg of meat after slaughter or death. At present, meat meal is manufactured in India on a small scale. The process of manufacturing meatmeal is very simple. First the bones and the meat are cooked or digested in a special containers for two to three hours. The bones are then separated from the meat. This meat is dried on a sandbath till it is brittle and then it is powdered. The drying can be done in double-jacket trays worked by steam and the material dried over the steam for an hour. Mutton squeezer can also be used for removing water content of the meat, and then dried. Meatmeal is a quick-acting manure and is effective for all crops, on all soil types. Its application is similar to oilcakes. Meatmeal contains about 10.5 per cent N and 2.5 per cent $P2O5$.

Bloodmeal

An adult cattle gives about 14.0 kg, and a goat or sheep about 1.40 kg of blood. Slaughter-houses should be provided with pucca or concrete floor with a central drain leading into a blood storage tank. The blood is first treated with commercial copper sulphate @ 125 gm per 100 kg of blood. It is then evaporated to dryness on a sand bath. Next, it is spread on a concrete floor covered by a net, and allowed to dry in sun. When completely dried, it is powdered, bagged and marketed as bloodmeal. Bloodmeal is a quick-acting manure and is effective for all crops on all soil types. It should be applied like oil-cakes. Bloodmeal contains 10 to 12 per cent of N and 1 to 2 per cent of P_2O_5.

Fishmeal

Non-edible fish carcasses are used to prepare fishmeal. The fish is dried, crushed or powdered and filled in bags. Fish manure is available either as dried fish, or as fishmeal, or powder. The manurial constituents present in it vary with the type of fish. Fishmeal is quick-acting organic manure and is suitable for application to all crops on all soils. It should preferably be powdered before use. Fishmeal contains 4 to 10 per cent N, 3 to 9 per cent P_2O_5 and 0.3 to 1.5 per cent K_2O

Horn-and-hoofmeal

A healthy animal gives about 3 to 4 kg of horn and hoof. These materials, are cooked dried and powdered before bagging. This meal contains about 13 per cent N.

AGRONOMIC IMPORTANCE OF ORGANIC MANURES

The agronomic value of an organic material is the increased crop yield or quality derived from its application. There is considerable published information which verifies the beneficial effect of organic materials on crop yield, but very little on crop quality. Crop yield response to the addition of organic materials is highly variable and is dependent upon the crop, soil type, climatic conditions, management system, and the organic material used. In most cases, crop yield response to the addition of organic materials is non-linear. Greatest yield response is obtained with the first few increments of organic material, followed by progressively smaller yield increases with additional increments. Thus, as with chemical fertilizers and other production inputs, crop yield response to an organic material follows the law of diminishing returns, resulting in decreased value per unit of material with increased application rates. Obviously, both agronomic and economic values per unit of organic material, to the farmer are correspondingly higher at low, rather than high, application rates. Reliable estimates of the economic value of organic materials depend upon the accurate assessment of agronomic data relating the crop yield response to the application of a particular material.

Organic Residues and Crop Response

Organic manures such as farmyard manure (FYM), rural and urban compost, green manure and crop residues are low in major plant nutrients as compared to chemical fertilizers. Those containing higher percentages of major plant nutrients are called concentrated organic manures such as oil cakes, blood and meat meals, guano and poultry manures. Well composted organic manure by traditional methods contains 0.5-1.0% N, 0.6% P2O5 and 0.5% K2O. Ten tons of compost will 50-100 kg N, 20 kg P2O5 and 50 kg K2O out of which 15 to 30 kg N, 15 kg P2O5 and 35 kg K2O will be available to crops during first year of application and the rest become available to subsequent crops. FYM and compost possess similar properties and their chemical contents vary depending on the nature of organic material used in its preparation.

Organic residues of plant and animal origin have been considered as a source of immense practical value for crop productivity. Organic residues including green manures, animal and municipal wastes serve as effective sources of plant nutrients and humus. Tropical and subtropical soils, as those in India, are generally deficient in organic matter and plant nutrients due to rapid loss of these components by biodegradation. To make up these losses, extensive utilisation of organic residues in agriculture can pay dividends provided suitable technologies to achieve these are available.

The relevant approaches to tackle the problem of organic manuring could be (a) improvement in the process of composting to upgrade the quality and nutrient status of organic manures and (b) utilisation of the available organic resources alongwith inexpensive low grade rock minerals in the natural plant production cycle.

The potential of crop residues/straw of some of the major cereals and pulses in India revealed that there is huge quantity of renewable crop residues produced in the country. The five major crops alone yield approximately 141.2 million tonnes of straw which at even most practical level will contribute about 0.7, 0.84 and 2.1 million tonnes of N, P_2O_5 and K_2O, respectively (Gaur *et al.*, 1984). Even if 50% of these crop residues are utilised as animal feed, the rest could be mobilised for recycling of plant nutrients.

Macronutrient Availability

Mineralization of plant nutrients and their availability to plants is one of the major consequence of organic recycling in soil. The quantity released depends mainly on the content of a particular nutrient in organic residues undergoing decomposition.

In general, there is immobilisation of nitrogen due to decomposition of cereal straw. However, with FYM/compost application no such effect is observed. The addition of FYM and cereal residues result in improvement of total soil nitrogen (Bhardwaj and Gaur, 1985).

The effect of organic matter in reducing the intensity of phosphate fixation by the soil sesquioxides and maintenance of soil fertility by use of organic manures alongwith superphosphate was advocated by Datta and Srivastava (1963). Organic matter applied to soil @ 1% proved more effective in increasing the available P from native as well as added P sources (Vyas and Motiramani, 1971). It was observed that both organic acids and the humus fraction of decomposing organic matter were more efficient in releasing phosphorus from rock phosphate and tricalcium phosphate and reducing phosphorus fixation in soil (Gaur, 1969; Pareek and Gaur, 1973).

The role of organic manures, particularly FYM in maintaining soil fertility is well documented. Studying cumulative, direct and residual effect of FYM in potato based sequences, Sharma *et al.* (1979) reported that FYM applied @ 44 kg P/ha was as effective as SSP. FYM was intermediate in building up available P status of soil as compared to 44 and 88 kg P/ha as SSP. In sub-humid lateritic soils, P use efficiency under rice-rice sequence was increased tremendously with FYM application. Information need to be generated on what amount of fertilizer P could be curtailed for organic manures under different cropping sequences.

Response of crops to organic manuring depends on degree of decomposition of organic residues, C : N ratio, time of application, soil characteristics etc.

Organic manures, which were perhaps the major source of plant nutrients in traditional agriculture, received less emphasis with the advent of high analysis chemical fertilizers. Without detracting from the fact that chemical fertilizers will continue to be main instruments for enhancing the pace or agricultural production, the recent research indicate that a judicious combination of organic manures and fertilizers can maintain the long term soil fertility and sustain higher levels of productivity in a much better way. Studies under AICARP (1983-87) brought about the fact that in case of maize-wheat, pearlmillet-wheat and sorghum-wheat sequences, manuring with FYM during *kharif* left a significant residual effect on succeeding *rabi* crops ensuring thereby some saving of N fertiliser. Residual effect of *kharif* manuring in case of rice-rice and rice-wheat sequences was however, not well marked and full recommended dose of N need to be applied during *rabi* season. Earlier studies also suggested the use of FYM during the wet season for higher efficiency (Mahapatra *et al.*, 1974). Based on long term experiments, Panda and Sahoo (1989) observed that in rice-rice sequence N use efficiency increased to the tune of 3 to 18% under NPK + FYM compared with NPK alone (Table 5.4).

Nitrogen deficiency in tropical and especially in Indian soils is so common that the responses of field crops to fertilizer N are observed throughout the region. While the direct effect of N on crops is most eye catching, the carry over effect is generally negligible owing to various path ways of its loss from the rhizosphere. Most of the research on N management, therefore, concentrates on minimising N losses and increasing N use efficiency. Inclusion of legumes in crop sequences and use of organic manures and nitrification inhibitors alongwith N fertilizers have been identified as some important agro-techniques to increase the efficiency of directly applied N and also the yield of the succeeding crops.

Studying cumulative, direct and residual effect of FYM in potato based sequences, Sharma *et al.* (1979) reported that FYM applied @ 44 kg P/ha was as effective as SSP. FYM was intermediate in building up available P status of soil as compared to 44 and 88 kg P/ha as SSP. In sub-humid lateritic soils, P use efficiency under rice-rice sequence was increased tremendously with FYM application.

TABLE 5.4: Mean nutrient use efficiency (% recovery) in rice-rice cropping system

Treatment	Red loam soil % use efficiency			Lateritic soil % use efficiency		
	N K-R	P K-R	K K-R	N K-R	P K-R	K K-R
100% N	22-27	-	-	27-30	-	-
100% NP	33-35	45-47	-	26-30	11-22	-
100% NPK	39-36	68-52	42-58	36-39	20-25	66-69
100% NPK + FYM	42-54	73-123	43-92	39-51	31-34	68-74

Source: Panda and Sahoo (1989); **K** = Kharif, **R** = Rabi

In acid Alfisols of Meghalaya, SSP amended with organic manures not only increased the yield of maize and rice, but showed significant beneficial effect on succeeding oilseed crops. In a organic manures enriched with SSP or SSP + rock phosphate mixture @ 28 kg P/ha enhanced the yield in the order: poultry manure> piggery manure> FYM. Information on quantification of the effect of FYM on P economy is quite scanty. Precise information need to be generated on what amount of fertilizer P could be curtailed for organic manures under different cropping sequences. The potassium fertility map for India prepared by Ghosh and Hasan (1976) showed 20% districts as low in available K, 42% as medium and 38% as high. If the map is revised on the basis of latest information, significant changes in above categories is likely to occur. In general, most of the areas in north-western region are considered adequate in available K. However, continuous high intensity cropping may lead to depletion of native K to such extent that crops show responses to K fertilisation. Data generated under Long Term Fertilizer Experiments showed contribution of different plant nutrients towards yield increase with continuous cropping (Table 5.5).

In multiple cropping systems, a good deal of information has been generated for potato based systems. The results indicate that K applied to potato can accure benefit to the following crops. In red soils of West Bengal, residual effect of 100 kg K_2O/ha applied to potato was significant on the succeeding crops of maize, green gram and jute.

TABLE 5.5: Changes in the input response pattern in a system requiring NPK + FYM for maximum yield (example of maize at Palampur)

Item	Period	
	1973-75	**1982-84**
Total grain response to NPK + FYM (kg/ha)	4825	5866
Percent contribution of N alone	51	0
Percent contribution of P (over N)	22	42
Percent contribution of K (over N P)	0	23
Percent contribution of FYM (over NPK)	27	35

TABLE 5.6: change in soil fertility after 2 cycles of cereal-cereal and cereal-legume cropping systems

Cropping systems	Total nutrients applied (kg/ha)	Total nutrients removed (kg/ha)	Nutrients from soil reserve (kg/ha)	Net change (kg/ha)	Total biomass (q/ha)
Maize (F)-rice-wheat					1154
Nitrogen	315	520	205	+3	
Phosphorus	-	93	93	-9	
Potassium	-	729	729	-108	
Maize (F)-rice-lentil					1101
Nitrogen	260	541	281	+8	
Phosphorus	17.3	91	73	+4	
Potassium	-	635	635	-80	
Cowpea (F)-rice-wheat					760
Nitrogen	225	492	267	+4	
Phosphorus	-	87	87	-5	
Potassium	-	619	619	-70	
Cowpea (F)-rice-lentil					676
Nitrogen	170	486	316	+8	
Phosphorus	17.5	79	61.5	+2	
Potassium	-	509	509	-73	

Source: George and Prasad (1989); **F** = Fodder
Initial values of mineralizable N, available P and available K were 187, 20 and 316 kg/ha, respectively.

Inclusion of green manure in sugarcane based systems helped maintaining higher levels of exchangeable K in the soil (Singh and Yadav, 1985). It is noticed that cropping systems (like fodder maize-rice-wheat) involving only cereals remove higher amounts of K from soil compared with the systems involving pulse crops (Table 5.6). Higher depletion of soil K under cereal based systems has also been recorded in other studies (Sharma and Prasad, 1974; Prasad and Prasad, 1983; Kemmler, 1986).

Micronutrient Availability

Sharma and Meelu (1975) reported that continuous application of organic manures improves the availability of Zn in soil and may not be sufficient to meet the requirements of the immediate crops. This is due to low Zn content of organic manures and slower mineralization particularly during rabi season (Tiwari et al., 1980b). This is why rabi crops like wheat show lower response to organic manures as micronutrient carriers. Mann et al (1978) observed that in case the organic manure is applied to a previous kharif crop, only then may enhance the yield of succeeding wheat crop. Though several single crop based results emphasise varying degree of effect of organic manures on micronutrients, information is still quite meagre.

Contrary to macronutrients, the range between deficiency and toxicity limits of micronutrients is quite narrow for most of the crops. Inclusion of micronutrients in fertilizer schedule should, therefore be advocated after careful appraisal through soil and plant tests.

Tissue analysis can also disclose the hidden hunger. The crux of induced deficiency is the nutrient imbalance manifested through their interactions. Mg deficiency is the classical example of induced one due to high levels of potash in many fruit and plantation crops. The strategy to rectify that would be reducing the K application to soil and foliar application of Mg.

Nitrogen Economy through Legume Crops

Several workers in India have reported the beneficial effect of legumes on nitrogen status of soil (Sadanandan and Mahapatra, 1973; Tiwari et al., 1980b; George and Prasad, 1989). The capacity of legumes to fix the atmospheric N can be exploited to meet partially the N requirement of crops by including them in cropping systems. The yield of wheat and maize was comparatively higher when preceeded with legumes. Nutrient uptake studies in exhaustive cereal-cereal cropping sequences indicated that even with application of recommended doses of fertilizers there is a deficit balance of N whereas the sequences involving legumes resulted in build-up of soil nitrogen (Rao and Sharma, 1978). They have recorded a net negative balance of 53 kg N/ha under maize-potato-wheat and an average gain of 81 kg N/ha under maize-potato-green gram at the end of two years. Inclusion of legumes in cropping systems improved organic carbon and mineralizable N content of soil (George and Prasad, 1989). Higher accumulation of organic carbon in soils consequent to inclusion of legumes in rice based cropping systems has also been noticed in Andamans (Mongia et al., 1989).

Nitrogen Economy through Green Manuring

Green manuring has proved a feasible and reliable practice to improve the productivity of important cropping systems like rice-wheat and soil fertility as well. Green manuring done during the months of May-June added 60-70 kg/ha of fresh biomass to the soil. At Pantnagar, increase in rice and wheat yields was largest in case of Sesbania compared with other green manures. As shown in (Table 5.7) beneficial effect of green manuring with Sesbania in an upland rice-wheat sequence has also been reported in alluvial soils of Kanpur (Tiwari et al., 1980a).

In this study green manuring plus 40 kg N/ha compared well with 120 kg N/ha alone showing thereby an economy of 80 kg N/ha. Transplanting rice just after ploughing down the green manure crop or after 15 days time allowed for decomposition could not make significant difference.

It is also suggested that in a rice-wheat cropping system, green manuring crop can be taken as a catch crop after the harvest of wheat for manuring the forth coming rice crop in normal as well as partially reclaimed saline-sodic soils (Tiwari *et al.*, 1979). Beneficial effect of cowpea green manuring and soil incorporation of cowpea residues, in rice-based cropping systems has been established (John *et al.*, 1989).

TABLE 5.7: Effect of different summer green manure crops on rice and wheat yield (kg/ha) in rice-wheat sequence.

Green Manure Crops	Yield (kg/ha)			
	Site I		Site II	
	Rice	Wheat	Rice	Wheat
Sesbania	6536	3570	7143	4118
Sunnhemp	5986	3346	6676	3918
Cowpea	5765	3202	6581	3809
Control	-	-	6135	3524
C.D. (0.05)	428	NS	220	320

Source: AICARP Status Paper (1983-87)

TABLE 5.8: Residual effect of Sesbania green-manuring on wheat

Treatment	Yield without N (q/ha)	Response to N at			
		N_{40}	N_{80}	N_{120}	Mean
	1975-76				
Fallow	16.75	0.35	2.05	6.00	18.85
Response*	-	0.9	2.6	5.0	-
Green manuring	23.10	4.70	8.70	10.40	29.05
Response	-	11.8	10.9	8.7	-
	GM	**N**	**GMXN**		
C.D. (0.05)	0.83	1.14	1.66		
	1976-77				
Fallow	7.17	1.08	2.48	5.555	9.45
Response	-	2.7	3.1	4.6	-
GM_1	13.65	2.25	7.37	11.57	18.95
Response	-	5.6	9.2	9.6	-
GM_2	13.87	1.87	6.48	12.55	19.09
Response	-	4.7	8.1	10.5	-
C.D. (0.05)	0.71	0.83	1.44		

Source: Tiwari et al. (1980a); *kg grain per kg N
GM_1 = Transplanting rice just after ploughing; GM_2 = Transplanting 15 days after ploughing

Cowpea residues gave higher agronomic efficiency as compared to green manure and urea-N under both upland and lowland conditions. These results indicate that in a rice-wheat sequence, green manuring in rice would be beneficial not only in increasing the yield of rice but also in enhancing the yield of wheat and improving the over all soil productivity.

Wheat

Experiments conducted at IARI, New Delhi during rabi 1987-88 showed that application of 5 t/ha straw was superior to control and 10 t/ha straw treatments. Both the microbial inoculants used singly or in combination with chemical nitrogen at 60 kg N/ha improved the grain yield appreciably over 60 kg N/ha alone and combination of both Azotobacter and P. striata with 60 kg N/ha proved to be the best treatment and recorded grain yield which was comparable to 120 kg N/ha treatment.

In another field experiment conducted at IARI during *rabi* 1988-89 it was observed (Table 5.9) that application of the microbial inoculants singly alongwith 60 kg N/ha resulted in a significant increase in grain yield over 60 kg N/ha alone. Application of 60 kg N/ha + Azotobacter gave significantly higher yields. Gaur *et al.* (1980) conducted field trials on the effect of organic materials and phosphate dissolving cultures on the yield of wheat. The results indicated that rock phosphate application significantly increases the yield of wheat crop in all its treatments except in paddy straw at C/N 40. The combined application of paddy straw, with nitrogen and rock phosphate together with phosphate dissolving culture gave the maximum yield of wheat crop. In another trial 47.4% higher yield was recorded over control where maize stubbles were applied together with nitrogen, rock phosphate and phosphate dissolving culture. The results indicate that an integrated use of organic and chemical fertilizers would be more beneficial. Verma and Mathur (1989) reported that application of single microbial inoculant in combination with 60 kg N/ha as urea and 60 kg P_2O_5/ha as rock phosphate in straw amended soil produced significant effect on yield of wheat. Combined inoculation of organisms had a repressive effect on growth and yield.

TABLE 5.9: Effect of fertilizer and microbial inoculant on wheat crop in rice straw amended soil

Treatments	Grain yield (kg/plot)		
	No straw	5t straw	10t straw
N60	2.22	1.78	2.15
N120	2.43	2.05	2.70
Azotobacter	1.18	1.30	1.08
P solubilizer	1.27	1.45	1.08
N60 + Azotobacter	2.68	2.22	2.38
N60 + P solubilizer	2.40	2.08	2.42
N60 + Azotobacter + P solubilizer	2.38	2.13	2.23
N60 + Pyrite	2.37	2.33	2.40
C D at 5%	0.25		

Source: Verma and Mathur (1989)

Maize

A positive effect of organic residue application on the yield of maize crop was observed by many workers. Singh and Verma (1968) observed a significant effect of application of FYM together with P fertilizer on the yield and P uptake by the maize crop. Mann et al. (1978) studied the effect of FYM or micronutrients in sandy loam soils. A positive response to foliar application of boron and molybdenum or to soil application of boron was observed in presence of FYM. Gaur and Mathur (1979) showed that significantly higher increase in yield was obtained when one third of organic matter as neem cake was incorporated alongwith straw, followed by FYM treatment. Straw application at 0-15 cm proved better than 0-7.5 cm depth.

Rice

Organic matter application in rice proved better than other inputs. A summary of 341 experiments on rice crop showed that response to application of 12.6 t of FYM or compost/ha varied from about 100 kg increased yield in Maharashtra to 216 kg increased yield/ha in Orissa (Gaur et al., 1984). Gupta and Sinha (1974) reported that application of ripened compost gave better rice yield in different soil types of Bihar. Rudraskha et al. (1978) observed that application of 75 t FYM/ha together with gypsum at 7 t/ha to alkali soil gave highest yield followed by FYM and half gypsum requirements. The results of a long term experiment (10 years) in sandy loam laterite soils of Bhubaneswar showed that application of FYM to supply 45 kg N together with 45 kg N from ammonium sulphate gave the highest grain yield of 2667 kg/ha, an increase of 520 kg/ha (Sahu and Nayak, 1971). In a pot and field trial with rice, application of undiluted or diluted cattle shed washings increased growth, dry matter accumulation, nitrogen uptake and rice yield (Jha et al., 1978).

Sugarcane

Results of 258 experiments conducted in eight states and 19 experimental stations in India showed average response of the order of 8 t sugarcane to the application of 25t of FYM or compost. Maraikulandi and Morachan (1964) reviewed the results of manurial trials on sugarcane conducted in Tamilnadu. A mixture of ammonium sulphate and groundnut cake was better than either groundnut cake or ammonium sulphate alone. Singh and Kanwar (1964) studied the manurial value of different oil cakes on sugarcane and observed that castor cake gave the maximum cane yield. Significant differences were not observed between various cakes, ammonium sulphate, blood meal and fish meal manure.

Crop Residues as Soil Mulches

The effect of crop residues on soil productivity and as surface mulches are quite appreciable. Mukherjee and Gaur (1984) reported that various mulched treatments increased ammoniacal and nitrate nitrogen, available phosphorus, humus content and microbial population over un–mulched control. Paddy straw as soil mulch significantly increased grain yields of different crops such as wheat, pea, green chickpea and maize over un–mulched treatments, (Gaur, 1978; Gaur and Mukherjee, 1980). Mukherjee and Gaur (1985) reported that application of FYM and green or to soil under mulched condition increased the grain yield as well as ammoniacal and nitrate nitrogen, available phosphorus and humus content of soil.

Economic Value of Organic Manures

The economic value of an organic waste or residue to a farmer is the value of the increase in crop yield and/or crop quality that is derived from its use. Since crop yield response to an organic amendment follows the law of diminishing returns, the average yield increase is always larger than the yield increase attributable to the incremental unit of organic material. If the price of the farmer's product is the same regardless of the quantity produced, which is usually the case, then the revenue derived from the average unit of organic material will be greater than the revenue from the incremental unit. Thus, the value of the average unit of organic material will be higher than the value of the incremental unit of organic material.

Farmers can be expected to utilize organic materials to the point where the revenue from the incremental unit is equal to its price, assuming application is included in the price. They would certainly use no more than that quantity since additional application of the organic material would yield a loss. Likewise, they would not wish to use less since total profits would decline. Thus, an estimate of the value of the incremental waste is necessary by management agencies when analyzing the marketability of a recycled product. However, public decision-makers should also be interested in the estimates of the total economic benefits from resource recycling alternatives, such as composting, compared with ultimate of the value of the average unit of organic material would be the appropriate figure to use in public project analysis.

CHAPTER 6

Use of Non-Traditional Additives for Organic Farming

A number of products are now available that are generally referred to as soil and plant additives, of non-traditional nature. These products include (i) microbial fertilizers and soil inoculants which are purported to contain unique and beneficial strains of soil microorganisms, (ii) microbial activators that supposedly contain special chemical formulations for increasing the numbers and activity of beneficial microorganisms in soil, (iii) soil conditioners that claim to create favourable soil physical and chemical conditions which result in increased growth and yield of crops, and (iv) vermicompost which helps in improving soil health and fertility.

BIOLOGICAL NITROGEN FIXATION

Nitrogen is the mineral element most commonly limiting in crop production. Despite its abundance in the atmosphere, plants are not able to assimilate the molecular form and so must satisfy their nutritional requirement from fertilizer or biologically fixed nitrogen, courtesy of a restricted number of diazotrophic bacteria. Because the manufacture of a kilogram of fertilizer nitrogen consumes nearly two cubic metres of natural gas, there is much concern that many countries will be unable to afford the energy expense of raising crop yields to levels commonly experienced in western agroecosystem. To meet global needs of 200 million metric tons of nitrogen in the next century could consume one-tenth of the earth's natural gas reserves, assuming that farmers could afford the price.

Currently, biological nitrogen fixation accounts for perhaps two-thirds of crop nitrogen needs worldwide. However, requirements in the industrialized nations are largely met by chemical fertilizers, whose production and use accounts for nearly half of all energy expended in the farming sector. In order to meet future demands, it will be necessary to drastically improve efficiencies of fertilizer production (perhaps with development of zero-direct energy input catalysts) and of biological nitrogen fixation.

Biological reduction of molecular nitrogen to ammonia is restricted to certain bacteria possessing the nitrogenase enzyme complex. Included are free-living species of *Klebsilla*, *Azotobacter*, *Clostridium*, *Azospirillum* (and other heterothophs), photosynthetic bacteria, cyanobacteria (blue-green algae), and bacteria that normally fix nitrogen only when symbiotically associated with a plant. Symbiotic nitrogen-fixation is accomplished in the *Azolla-Anabaena* association, by actinomycetes in the genus *Frankia* (which nodulate Alnus and Comptonia species), and by *Rhizobium*. Nodulation of legumes by rhizobia is the most important source of biologically fixed nitrogen in agriculture and it affords the greatest potential for genetic improvement in the short to near long term.

Until recently, legume-*Rhizobium* studies had been confined to temperate crops. Potential for increasing input of biological nitrogen via the legume-*Rhizobium* symbiosis will be realized along two avenues. First will be development of recently discovered legume species and their contribution to tropical agriculture. Second will be improving the temperate legume-*Rhizobium* symbiosis by genetic engineering of more energy efficient and more competitive rhizobial strains.

Biological nitrogen fixation is as energy expensive to the legume as is fertilizer nitrogen to the farmer. This summarizes a great number of biological and technical limitations to what can be expected of the legume-*Rhizobium* symbiosis. Biochemical limitations include the large size and complexity of nitrogenase itself, its sensitivity to oxygen and temperature, a sluggish turnover rate, the requirement for special electron donors and large amounts of ATP, inhibition by fixed nitrogen, and its substrate promiscuity (particularly the production of H_2 from H which is an extremely wasteful reaction).

Because of the large carbohydrate requirement for nodulation and nitrogenase activity (up to 18 kg/kg N), the major limitation to the overall process is the inefficiency of photosynthesis. For example, soybean yields of 1800 kg/ha represent a conversion efficiency of solar energy of only 0.1 to 0.2% (efficiencies in cereals seldom reach 4% of available sunlight). Nevertheless, engineering *Rhizobium* for efficiency of competition in the rhizosphere, nodulation, and nitrogen fixation offers much hope in light of tremendous strides in our understanding of the genetic and biochemical complexities of the symbiosis.

Nitrogen fixing organisms can be provided to the farmers in the name of microbial inoculants otherwise termed as Biofertilizers. Broadly, these are the preparations containing live or latent cells of efficient strains of beneficial microorganisms used for application to seed or soil with the objective of increasing their numbers and accelerate certain microbial processes to augment the extent of the availability of nutrients in an assimilated form by the plants. The biofertilizers containing biological nitrogen fixing organisms are of utmost importance in agriculture in view of the following advantages:

1. They help in the establishment and growth of crop plants and trees
2. They enhance biomass production and grain yields by 10-20%
3. They are useful in sustainable agriculture
4. They are suitable in organic farming
5. They play an important role in agroforestry / silvipastoral systems

Types of Biofertilizers

Rhizobium

The most widely used biofertilizer is Rhizobium which colonizes the roots of specific legumes to form tumour like growths called root nodules. It is these nodules that act as factories of ammonia production. The Rhizobium-legume association can fix upto 100-300 kg N/ha in one crop season and in certain situations can leave behind substantial nitrogen for the following crop. The range of nitrogen fixed per hectare per year by different legumes is 100-150 kg for

clover, 80-85 kg for cowpea, 100-300 kg for alfalfa, 240-320 kg for faba bean, 90-100 kg for lentil, 150-200 kg for lupins, 50-60 kg for groundnut, 60-80 kg for soyabean, 50-55 kg for mungbean and 100-400 kg for pasture legumes.

Azotobacter

The beneficial effects of Azotobacter biofertilizer on cereals, millets, vegetables, cotton and sugarcane under both irrigated and rainfed field conditions have been substantiated and documented (Pandey and Sushil Kumar, 1989). Application of azotobacter has been found to increase the yields of wheat, rice, maize, pearl millet and sorghum by 0-30% over control. Apart from nitrogen, this organism is also capable of producing antibacterial and antifungal compounds, hormones and siderophores.

Azospirillum

Certain micro-organisms like bacteria and blue-green algae have the ability to use atmospheric nitrogen and transport this nutrient to the crop plants. Some of these nitrogen fixers like rhizobia are obligate symbiont in leguminous plants, while others colonize the root zones and fix nitrogen in loose association with plants. A very important bacterium of the latter category is Azospirillum which was discovered by a Brazilian scientist and which attracted head-lines in the mid 1970s. The crops which respond to Azospirillum inoculation are maize, barley, oats, sorghum, pearl millet and forage crops. Azospirillum applications increase grain productivity of cereals by 5-20%, of millets by 30% and of fodder by over 50%.

Blue-Green Algae

The utilization of blue-green algae as a biofertilizer for rice is very promising. A judicious use of these algae could provide to the country's entire rice acreage as much nitrogen as obtained from 15-17 lakh tonnes of urea. Methods have been developed for mass production of algal biofertilizer and rice growers in many parts of the world have already started using these algae (Venkataraman, 1981, 1988). Recent researches have shown that algae also help to reduce soil alkalinity and this opens up possibilities for bioreclamation of such inhospitable environments.

Azolla

A small floating water fern, Azolla is commonly seen in low land fields and in shallow fresh water bodies. This fern harbours blue - green algae, Anabaena azollae. The Azolla-Anabaena association is a live floating nitrogen factory using energy from photosynthesis to fix atmospheric nitrogen amounting to 100-150 kg N/ha/year from about 40-60 tonnes of biomass (Hamdi, 1982; Singh, 1988). Reports of its current use as a biofertilizer for rice in China, Vietnam, India, Indonesia, Thailand and other East and South Asian countries are available. An integrated system of rice Azolla fish has been developed in China.

Mycorrhizae

Mycorrhizae are the symbiotic association of fungi with roots of vascular plants. The main advantage of mycorrhizae to the host plants lies in the extension of the penetration zone of the root fungus system in the soil, facilitating an increased phosphorus uptake. The interconnected net-work of external hyphae acts as an additional catchment and absorbing surface in the soil beyond the depletion zone that would otherwise be inaccessible to the plant roots. Endotrophic mycorrhizae have been shown to be present in a wide range of horticultural species including apple, walnut, almonds, citrus, avocado, strawberry and grape (Mosse, 1963; Harley, 1969; Possingham and Obbink, 1971). In many cases the mycorrhizae have been shown to markedly improve the growth of plants.

In India the beneficial effects of vasicular-arbuscular mycorrhizae (VAM) have been observed in fruit crops like citrus, papaya and *litchi*. Inoculation with *Glomus fasciculatum* fungus was found to increase the dry matter content and the uptake of Fe, Zn, and Mn by different rootstocks of citrus (Manjunath *et al.*, 1983).

In litchi, positive correlations of VAM population in the rhizosphere of soil and mycorrhizal infection in the roots with leaf area and fruit have been noticed (Rana and Srivastava, 1984). In papaya, inoculation with *Glomus fasciculatum* and *Glomus mossae* was found to increase the plant height, dry matter content and nutrient uptake with lower doses of applied phosphorus (Sukhada, 1988).

Recent studies have shown the possibility of domesticating *Mycorrhizae* in agricultural systems (Hayman, 1982; Tilak, 1987). Mycorrhizal fungi's assist in the uptake of phosphate (Tinker, 1980) and trace metals (Tinker, 1978; Timmer and Leyden, 1980) and possibly influence water and nutrient via hormonal influences (Allen *et al.*, 1982) is not in doubt. Lack of a suitable inoculum production technology is the major limitation for commercial exploitation of this system.

TABLE 6.1: Effect of mycorrhizal (VAM) inoculation on crop yields

Crop	P-fertilizer supplementation (kg P_2O_5/ha)	% yield increase over control
Finger millet	19	18
Soybean	25-50	19
Chillies	37.5	55
Chickpea	40	25
Groundnut	-	10-20

Source: Venkatraman and Tilak (1990)

Cost Effectiveness

Production of chemical fertilizer nitrogen is a high energy-budgeted process based on fossil fuels with one unit of nitrogen requiring two units of fuel. The energy requirement for one kg fertilizer is 80 MJ or 11.2 KWH for nitrogen. 12 MJ or 1.1 KWH for phosphorus and 8 MJ or 1 KWH for potash.

For example, our earth receives over Rs. 580 thousand million worth of nitrogen every year through biological sources. In terms of crop productivity, for three legumes, chickpea, lentil and soybean alone, rhizobial inoculation can generate an additional gain of approximately Rs. 11,800 million from the entire cropped area, for which the inoculant cost will be only around Rs. 35-40 million. In the case of algae, the cost of inoculant 10 kg/ha will be around Rs. 20 and its nitrogen contribution in terms of chemical fertilizers is around Rs. 220-400/ha. Further, if rural production is encouraged, it has a potential of generating an additional income to the farmers of about Rs. 1000-1500 from half an acre of land through the sale of the produce. Under Indian conditions crops like groundnut, soybean, chickpea and finger millet have shown 10-25% increase in their gram yields. Chillies were found to be more responsive with 55% increase in yield registered in its favour.

TABLE 6.2: Benefits derived from the use of biofertilizers in India

Biofertilizers and crops	Grain Yield (kg/ha)		Increase	Increase in yield (%)	Monetary benefits/ha for an input costing Rs 10-20/ha
	Control	Inoculated	(kg/ha)		
Rhizobium Chickpea	1956	2228	272	13.9	1101.60
Rhizobium Pigeonpea	1985	2182	197	9.9	1280.50
Blue-green algae Rice	4175	4650	475	11.4	1311.00
Azolla–Rice	2800	3480	680	24.3	1876.80
Azospirillum Sorghum	3130	3700	570	18.2	598.00
Azospirillum Pearl millet	1430	1789	359	25.1	589.80
Azotobacter Cotton	1254	1339	85	6.8	433.50

Source: Venkatraman and Tilak (1990)
Calculated on the basis of wholesale price quoted in Agricultural situation in India, 1984.

Biofertilizers and Sustainable Agriculture

Biofertilizers have for long witnessed shifting fortunes in agriculture. While the technology of using them has been successfully exploited in many developed countries, competent exploitation in developing countries has been hampered by several factors, including lack of trained personnel, lack of appreciation of the benefits of inoculation and absence of suitable industrial support. A major problem, especially relevant to our biofertilizer programme has to do with relations between research institutions and industry. A transfer of technology between the research institutions and industry is essential. It must be promoted as effectively as possible through cooperation between institutional research, which is usually public and industrial research and production, which are either private or public. At the same time, we should not overlook the significance of ensuring and maintaining a high quality standard of the product. The limited shelf-life, particularly of bacterial biofertilizers dictates that product streams must be handled with a quick delivery system at low temperatures. A strong extension and training programme, actively supported by research and industry is the need of the hour. We should recognize that in adopting a rational approach to the use and management of natural resources in sustainable agriculture, the microbial fertilizers hold vast potential for the future.

In developing countries, the most important challenge is to produce sufficient food for the growing population from inelastic land area. Products of biological origin can be advantageously blended to replace a part of the energy-intensive inputs. It is in this context, biofertilizers can provide to the small and marginal farmers an economically viable lever for realizing the ultimate goal of increasing productivity. These microbial systems siphon out appreciable amounts of nitrogen from the atmospheric reservoir and enrich the soil with this important but scarce nutrient. The crop-microbial-soil ecosystem can, therefore, be energised in sustainable agriculture with considerable ecological stability and environmental quality. Although the potential of biofertilizers in crop production systems has been well documented and substantiated, the major reason for the shifting fortunes of these biological inputs lies in the lack of an organised industrial back-up, an effective quality control system, powerful extension machinery and a broad research base.

Research Priorities

The future thrust in biological nitrogen fixation could well be the development of self fertilising crops that can meet their nitrogen needs from air (Venkatraman and Tilak, 1990). The recent biotechnological and genetic engineering capabilities have generated considerable optimism in this direction. Scientists have already started looking into the molecular architecture of the genes responsible for fixing nitrogen with a view to transfer them between different systems. The current approaches are directed towards developing super-strains that can nodulate a wide range of leguminous plants; transferring the nitrogen fixing, nodule specific genes from legumes to non-legumes through protoplast fusion; making nitrogen fixing bacteria such as Rhizobia infect non-leguminous plants and fix nitrogen and transferring the genes for nitrogen fixation (nif gene) from bacteria that have it to non-nitrogen fixing ones. A detailed genetic map of the nif region is now available and the nif gene cluster has been cloned, amplified and transferred to other bacteria and even to yeast. Although transfer was successful, expression did not occur and so far no successful expression of nif genes has been reported in a higher plant. Nevertheless, studies are continuing. Many commercial companies are actively engaged in research in these areas. In blue-green algae, biotechnology shall aid in having strains with high rates of nitrogen fixing ability and a high tolerance level to environmental stress. Current research in Azolla is directed towards protoplast fusion, hybridization and induction of sporulation.

SOIL CONDITIONERS

Soil conditioners are termed as materials which when added to the soil help in improving or maintaining its physical conditions. For centuries, farmers world wide know the importance of crop residues, organic manures and other substances of organic nature in improving the physical conditions of these soils. These materials help in increasing the infiltration of water and reducing its run off, binding soil particles especially in sandy soils, nutrient availability and plant growth and development.

The formation and stability of soil aggregates is a very important phenomenon. Although there is still some uncertainty about the exact mechanism which lead to formation of granules, several factors are responsible for their geneses. These include- I) soil tillage, ii) wetting and drying, iii) freezing and thawing, iv) activities of plant roots and soil fauna, v) decomposition of organic matter and vi) the modifying effects of adsorbed cations.

The stability of aggregates is of great physical importance. Three major factors influence aggregate stability- I) binding action of soil microorganisms, ii) cementing action of intermediate products of microbial synthesis and decay, such as gums and certain polysaccharides, and iii) cementing action of the more resistant stable humus components, which contributes to long term stability of the soil aggregates. Aggregate stability is not entirely an organic phenomenon as interaction does take place between organic and inorganic components. Polyvalent inorganic cations (*e.g.*, Ca^{++}, Mg^{++}, Fe^{++}, and Al^{3+}) are thought to act as bridges between the organic matter and soil-clays, thus encouraging development of clay-organic matter complexes. Soils rich in kaotinite and hydrons oxide clays have aggregates with highest stability. The larger the aggregates in a soil, the lower is its stability.

Now, synthetic organic materials are available which help in formation and stabilization of soil aggregates. These materials are polyelectrolytes, including polyvinylites, polyacrylates, cellulose gums, lignin derivatives and silicates. These materials improve the physical conditions of the soil alongwith its structure. These conditioners have been found to be useful in checking soil erosion especially in slopes. These materials are becoming popular with growers of high value crops, especially in the developed countries.

VERMICOMPOST

It is a method of making compost with the use of earthworms which generally live in soil, eat biomass and excrete it in digested form. This compost is generally called vermicompost or wormicompost. It is estimated that 1800 worms which is an ideal population for one sq. meter can feed on 80 tonnes of humus per year. To prepare an ideal vermicompost, the following procedure is adopted.

Each shed measuring 20 ft x 80 ft is to be constructed with the help of locally available material like bamboos, stems of trees etc. A hut type structure is built with the help of these articles. The roof is made from dried grass, typha leaves, bamboo sticks, etc., in such a way that the hut may be protected from rain water and sun heat. Each hut may accommodate at least four vermibeds of 3 ft. width.

These beds are prepared by putting 2 to 3 cm thick layer of farm manure as first layer followed by 10 to 15 cm of biomass with 200-250 worms per sq. ft. collected locally may be added and the bed should be kept sufficiently moist. This layer should be followed by a layer of 10 to 15 cm of half digested cow-dung layer which should be covered by a layer of leaves, trash etc., and water is sprinkled on the entire bed. The bed may be covered with palm leaves or coconut leaves or with any indigenous material. The pits should be kept constantly moist but never flooded.

A month later, the covered leaves should be removed and layers of organic waste not exceeding 6-7 cm should be added every alternate day. Watering should continue with each filling. When the pit is nearly full to a height of one metre, the material should be turned to provide aeration. After a month, the heap will be ready for harvest with good quality vermicompost. The dug out vermicompost should be heaped in an open place. The worms will find way to the bottom of the heap. The vermicompost from the top can be removed, dried and sieved for application in the field. The compost can also be enriched with micro-nutrients, bacteria etc. by adding them externally. 16 tonnes of compost can be obtained from 4 beds in 30 days after four months period of gestation.

AQUATIC WEEDS

Aquatic weeds are still regarded by many people as a menace, because they are not yet aware of the great potential and economic value of these profusely growing, un-controllable plants. These highly productive plants are generally more productive then conventional field crops. Moreover, when grown on wastes, they do not compete with conventional crops for fertilizer, water or land. Their natural profuse growth in the humid tropical and subtropical areas of the world, requiring no intensive cultivation, makes them a promising source of multipurpose raw material. As a matter of fact, these plants have proved to be good source of livestock feed, human food, fish feed, organic or biofertilizer, energy, fibre and paper. Furthermore, the aquatic weeds have the capacity to purify waste water through the uptake of dissolved nitrogen, phosphorus and undesirable excessive minerals, including heavy metals.

Organic Manures

The beneficial role of organic manures in maintenance of soil fertility and crop productivity has already been discussed before and thus hardly needs any emphasis. Organic manures act in many ways in augmenting crop growth and soil productivity. The direct effect of organics relates to the uptake of humic substances or its decomposition products affecting favourably the growth and yield of plants. Indirectly, it augments the beneficial soil microorganisms and their activities and thus increases the availability of major and minor plant nutrients.

Aquatic weeds have been reported to be a good source of organic manure. Many aquatic weeds contain appreciable amounts of nitrogen, phosphorus and potassium (Table 6.3). Composted aquatic weeds, when applied to farm land, have been reported to benefit the field crops (Table 6.4).

TABLE 6.3: Nutrient content of aquatic weeds

Weed	N (%)	P (%)	K (%)
Azolla pinnata	2.33-3.83	0.196	0.153
Eichhornia crassipes (small)	1.50-2.20	0.37-0.56	2.68-3.45
E. crassipes (large)	1.70-2.15	0.38-0.56	2.68-3.45
Lemna minor	3.56-3.75	0.23	0.18
Monochoria vaginalis	3.12-3.75	1.84-1.97	4.34-4.91
Pistia stratiotes	2.12-2.21	0.75-0.89	2.36-2.39
Salvinia sp.	1.12-2.05	0.34-0.61	2.10-3.30
Spirodela polyrhiza	2.76-4.54	0.37	0.59
Spirogela and Lemna	4.06	0.78	1.96

Source: Majid, 1983.

Table 6.4: Comparative effect of compost of aquatic weeds/algae, cow manure on crop yield.

Crop	Comparison of Manure	Increase in Yield
Rice	Aquatic weed (*Eichhornia*, *Pistia*, *Spirodela* mixed) Vs. cow manure	28.57% more in cow manure
Rice	Aquatic weed (*Eichhornia crassipes*) Vs. cow manure	22% more in aquatic weed
Rice	Alga (*Rhizoclonium*) Vs. cow manure	40% more in alga
Rice	Aquatic weed (*Spirodela*) Vs. cow manure	33.33% more in aquatic weed

Crop	Comparison of Manure	Increase in Yield
Wheat	Aquatic weed (*Eichornia crassipes*) Vs. cow manure	30.23% more in aquatic weed
Maize	Aquatic weed (*Eichhornia*, *Pistia*, *Spirodela* mixed) Vs. cow manure	40% more in cow manure
Soybean	Alga (*Spirogyra*) Vs. cow manure	55% more in the alga
Sesame	Aquatic weed (*Eichornia crassipes*) Vs. cow manure	21.5% more in aquatic weed
Sesame	Aquatic weed (Azolla) Vs. cow manure	50% more in aquatic weed
Sesame	Alga (*Spirogyra*) Vs. cow manure	37% more in the alga
Onion	Aquatic weed (*Eichhornia*, *Pistia*, *Spirodela* mixed) Vs. cow manure	13% more in aquatic weed
Onion	Aquatic weed (*Eichornia crassipes*) Vs. cow manure	25.5% more in aquatic weed
Basella rubra (Beng. Puin Shake)	Aquatic weed (*Eichhornia*, *Pistia*, *Spirodela* mixed) Vs. cow manure	9% more in cow manure
Spinach	Aquatic weed (*Eichornia crassipes*, *Pistia* mixed) Vs. cow manure	20% more in cow manure
Lettuce	Aquatic weed (*Eichornia crassipes*) Vs. cow manure	23.65% more in cow manure
Cabbage	Aquatic weed (*Eichornia crassipes*) Vs. cow manure	59% more in cow manure
Cauliflower	Aquatic weed (*Eichornia crassipes*) Vs. cow manure	35.62% more in cow manure
Cauliflower	Drain algae (*Phacus*, *Chlamydomonas*, *Oscillatoria*, Diatoms) Vs. cow manure	4.33% more in cow manure
Gourd	Aquatic weed (*Eichornia crassipes*) Vs. cow manure	55% more in aquatic weed
Beet	Drain algae (*Phacus*, *Chlamydomonas*, *Oscillatoria*, Diatoms) Vs. cow manure	7.34% more in cow manure
Turnip	Aquatic weed (*Eichornia crassipes*) Vs. cow manure	7.5% in cow manure
Radish	Aquatic weed (*Eichornia crassipes*) Vs. cow manure	3.70% more in aquatic weed
Brinjal (Egg plant)	Alga (*Spirogyra*) Vs. cow manure	2.71% more in the alga

Source: Majid et al. (1980)

Aquatic weeds not only provide crops with nutrients, they also improve soil's physical properties like, texture and water holding capacity of soil, which are very important in sandy, lateritic and heavy clay soils widely distributed in developing countries. Azolla, a good source of biofertilizer, is capable of reducing nitrogen demand of soil. Residual effect is also appreciable when aquatic weed compost is added to the soil.

Whereas organic manure has been found to be absolutely essential for maintenance of soil productivity, continuous use of chemical fertilizer, without the supplement of organic manure, has been reported to adversely affect the physical properties of soil. Scientists have been urged to find organic and biological techniques to replace excessive use of chemical fertilizers and pesticides in view of the problems of ecological imbalance.

Animal and Fish Feed

Aquatic weeds are a good source of protein (Table 6.5) and may be incorporated in the diet of livestock. Water buffaloes consume aquatic plants, such as, water hyacinth. This weed is also consumed by cattle, but for good consumption, it has to be mixed in right proportion with other feeds.

Ducks, geese and swans are commonly known to be herbivorous, although they also consume insects, snails etc. Water weeds may serve as a feed for those herbivorous water-birds that provide man with meat and eggs. In developing countries they are promising for the use in small farms.

Tilapia, Silver carp, South American silver dollar fish, Cray fish, Grass carp etc. are herbivorous fishes that feed on aquatic weeds. Grass carp is a quick growing fish that lives on water plants. Its flesh is highly prized in China. The fish prefers succulent submerged weeds which are difficult to control by conventional methods. Grass carp is native to cool-water rivers of China, but it also thrives in warm tropical waters. Cray fish is cultured on a large scale in rice fields in USA where they feed on aquatic weeds, aquatic insects, and rice stubbles.

TABLE 6.5: Protein content of important aquatic weeds

Weed	Moisture (%)	Total Solid (%)	Protein (dry basis) (%)
Azolla pinnata	88-91	9-12	23.03
Eichhorinia crassipes (small)	91.2-95	5-8.8	9.37-14.80
Eichhorinia crassipes (large)	91-95	5-9	9.14-14.37
Lemna minor	93.85	6.15	20.31-23.45
Monochoria vaginalis	93	7	18.75-22.50
Pistia stratiotes	95	5	12.81-13.27
Salvinia cuculata	97.4	2.6	12.8
Salvinia natans	94.87	5.13	9.37
Spirodela polyrhiza	95	5	17.25-28.39
Wolffia arrhiza	91.24	8.76	14.87-25.31
Combination			
Lemna, Pistia, Spirodela	92-94	6.8	26.5
Lemna, Spirodela	92.7	7.3	25.37-28.43

Source: Majid, 1983.

Source of Energy

Many developing nations like India, Bangladesh, Pakistan, Philippines and Thailand etc., have an inexhaustible supply of aquatic weeds which can form a potential source of energy. Fuel from aquatic weeds, even if in a small quantity has economic implications, especially for rural areas in developing countries. Aquatic weeds may be used for biogas production as an organic manure and soil conditioner.

Waste Purifier

Compounds containing nitrogen and phosphorus, the major ingredients in fertilizer, happen to be common pollutants in waterways. Raw sewage and sludge from domestic and industrial source contains various organic and inorganic materials, which may pollute the environment. These contaminants if not removed are quite hazardous from the point of human health. It has been reported (National Academy of Sciences, U.S.A., 1976) that some aquatic weeds can scavenge inorganic and some organic compounds from water. These weeds are capable of absorbing and incorporating the dissolved materials into their own structure.

Effluent, purified by aquatic weeds, is cleaned of its pollutants, and hence when released into waterways, causes less public health and environmental problems. The clean water produced, is generally suitable for re-use in irrigation and industry. Moreover, the aquatic plants can be harvested and safely used as manure, or for fibre or biogas production etc.

Aquatic weeds may also be used for removing potentially harmful elements or compounds from drinking water such as, cadmium, nickel, mercury, phenol, potential carcinogens etc. The aquatic weeds can remove and concentrate these harmful elements, which may become 4,000-20,000 times more concentrated in the plants than in the water. However, not all aquatic weeds are equally adapted for thriving on waste waters. Those that grow best include common reed, bulrush, water hyacinth, duck weeds, elodea, hydrilla etc.

Methanol as Fertilizer

Since methanol has long been known to be toxic to plants, its positive effects might appear to be unexpected. Recently, Nonomura (a Japanese-American farmer of Arizona) has discovered that spraying plants with a 20% solution of methanol, stimulates their growth. Nonomura, noticed that cotton plants wilted during the midday heat. Such wilting is caused by the more rapid loss of water from the leaves than its replacement by water taken up by the roots from the soil. Wilting causes the stomata to close and reduces photosynthesis because no new CO2 can enter the leaves. At the same time, photorespiration becomes more pronounced because of the lower CO2 level inside the leaves. Nonomura applied a dilute solution of methanol to the wilted cotton plants. Within hours, the sprayed plants were once again turgid, their stomata were open, photosynthesis resumed at high levels, and photorespiration went down. The growth rate also increased significantly, and the plants produced cotton bolls two weeks earlier than usual, thereby reducing their overall need for water. Such studies have also been conducted on cabbage plants, resulting in heads twice as large as those on untreated plants grown for the same length of time. Methanol also stimulates the growth rates of wheat and barley, and this should result in increased yields.

The discovery of this effect of methanol is so recent that almost no research has been done so far to understand the underlying mechanism.

CHAPTER 7

Agricultural Waste Management and Crop Production

In view of the energy and power crises besides the high cost of agricultural inputs such as fertilizers, pesticides and irrigation water, agricultural wastes are now considered quite an important component of farming especially that based on organic systems. Agricultural wastes can now be considered to include crop after harvest and primary processing, tree residues organic/plant residues from social forestry, animal excreta and processing left over from the slaughter houses and agro-industrial wastes. Thus, agricultural wastes comprise all organic wastes produced and disposed of or used in primary agricultural production. These wastes are exclusively animal manures and crop residues or mixtures of the two. Of course, these materials are essentially by-products of production rather than wastes in the strict sense of the word.

Important changes in agricultural production have been taking place over the past half century or so, and have had an important influence on the use of agricultural wastes as fertilizers. Changes in the availability of energy or in the price or availability of fossil fuels could have pronounced effects on the world food supply and how it is produced. Food supply in the developed countries could be particularly vulnerable if fossil fuel or an alternative is not available. It would take many years to get a skilled labour force and farm animals to cultivate the soil to produce food for the population with the traditional methods that were used up a few decades ago.

Plant Nutrition

Over the past century, many of the basic scientific principles of plant nutrition and soil fertility have been explained. In developed countries, inorganic chemical fertilizers have been widely accepted as a major source of improving and maintaining soil fertility.

With the advent of abundant and relatively inexpensive chemical fertilizers, agricultural manures became agricultural wastes, and what was formerly an asset is now often regarded as a disposal problem. In addition to these events, rapid and radical changes have taken place in methods of livestock production, particularly over the last two decades. Traditionally, cattle, pigs, and poultry were fed on free range or housed in relatively small numbers. Today, in most developed countries, intensive animal production is now being practiced where increasing numbers of livestock are being fed on fewer larger farms.

These changes have been accompanied by dramatic increases in the use of chemical fertilizers. In the region of 108 million tonnes of plant nutrients were used as chemical fertilizer worldwide in 1992-93 (FAI, 1994).

It is clear that the fertilizer value of agricultural wastes has a greater potential for agricultural production if properly used and this is particularly true for developing countries (Flaig *et al*, 1978).

Collection and Storage

Traditionally, animals were bedded with straw or other litter and the manure was mixture of urine, faeces, and straw. This farmyard manure was normally stored and spread as a solid. A high proportion of agricultural waste is still handled with straw, as farmyard manure. However, an increasing proportion is being stored and handled as a liquid or slurry where no bedding is used. This is particularly true on most new and large intensive animal enterprises.

Animal wastes can be divided into three main categories:

1. Farmyard manure, where straw or other bedding is used;
2. Slurry, which is a mixture of faeces and urine, often containing variable quantities of rainwater and wash water; and

3. Separated manure, where the waste is separated into two fractions-solid and liquid-which are handled separately.

The third category is receiving increased attention with the use of mechanical slurry separators, where the liquid fraction is spread or irrigated to land, and the solid fraction is handled separately as a solid. Other categories of manure in which non-mechanical separation takes place can be included here. In arid or semi-arid regions, manure quickly dries out by evaporation after it is produced in feedlots or corrals. This manure is handled as a solid and is sometimes stacked. With this method of storage there is no liquid for disposal; however, there can be runoff problems after heavy rainfall. In temperate regions with relatively high rainfall, such as Northern Europe, slurry is sometime pushed into compounds and solid/liquid separation takes place during storage. A solid crust forms on the top of the compound and sludge on the bottom with liquid in the centre (Robertson, 1977). This method is sometimes described as dungstead storage and is essentially non-mechanical solid/liquid separation. The liquid fraction can be drawn off and spread first, and then the solid fraction remaining can be spread.

Crop residues could be considered a source of agricultural wastes. Residues from most root crops, however, are returned directly to the soil during harvesting (for example, potato haulms). These residues could not be considered as wastes and are really in the same category as plant roots. The residues do, however, supply considerable quantities of essential nutrients for the succeeding crops after mineralizatin by soil micro-organisms.

Straw on soil after harvesting cereal crops is in a somewhat different category, as it can interfere with subsequent mechanical cultivation. To avoid this problem, straw is often burned or mechanically chopped into small pieces after harvesting. Straw left on the soil in this manner supplies the small quantities of phosphorus, potassium, and other nutrients present for subsequent crops. Due to the high carbon to nitrogen ratio of straw, the mineralizing organisms may reduce the available soil nitrogen, and extra should usually be applied-about 30 kg N/ha, if straw is not burned or removed. In recent years, straw is increasingly being used as source of feed for cattle. Research has shown that the feeding value and digestibility of straw can be increased by treatment with an alkali, such as sodium hydroxide, or by mechanical treatment, such as very fine grinding

Dairy cattle are sometimes housed in straw bedded sheds. The manure is usually allowed to accumulate to a depth of about 1 metre before cleaning. The manure is then either spread directly to land or stored in a heap for subsequent spreading.

Increasingly, manure is being collected and stored as slurry. The concept of handling manure as slurry containing a mixture of faeces, urine, and water is not a new one. Different methods of slurry handling have been practised over the past century.

Modern methods of handling animal wastes as slurry were greatly influenced by the development, in Scandinavia, of slatted floors for animal rearing. This development has revolutionized cattle and pig housing over the past three decades. Along the same lines is the "slope slot flush," which was developed and is currently studied in the United States. With this system, manure passes through slots into pipes in the floor and flows to storage tanks outside the building.

Free stall cubicle houses are a popular method of housing cattle, particularly dairy cows. With this system, the manure is usually cleaned by tractor scraper off concrete yards and passes into storage tanks. The storage tanks may be underground or overground and can be concrete or steel, glass lined, or made from other materials. Tractor-driven pumps are sometimes used to transfer the manure to an overground storage tank.

Slatted floors are extensively used for beef animals, and in the past few years, slatted passages for dairy cows have been widely adopted. With the slatted floor house, the manure is usually stored in 2-metre deep tanks beneath the house. There is usually adequate manure storage for the winter feeding period, and there is no labour involved in manure handling while animals are in the house. It passes directly as it is produced through the floor into the storage tanks below.

For pigs, straw is still sometimes used for bedding, but the majority of intensive units collects and store manure as a slurry. Part of the floor area is normally slatted and the slurry flows in manure channels to outside storage.

In the case of poultry, broilers are usually reared on deep litter. The litter is usually wood shavings, chopped straw, or similar materials. The house is cleaned out after about 11 weeks, when the broilers are sold. Fresh litter is then put in for a new batch of chicks. This manure usually has about 75% dry matter and is either stored in a heap or spread straight from the house. Laying hens are now usually housed in cages and the droppings pass through the cage floor. The droppings are usually taken by conveyor belt to a spreader or storage outside the house. The hen droppings have about 25% dry matter. Of all agricultural wastes, poultry manure contains the highest level of plant nutrients and therefore has the highest fertilizer value.

Many machines for spreading manure have been designed for disposal, so that manure is often spread unevenly. To get maximum fertilizer value from manure it is important that machines should be capable of spreading evenly, as with fertilizer distributors. Most manure is spread on the soil surface. However, the possibility of injecting slurry into the soil is receiving attention. Slurry injection into cultivated soil has been used for many years, and more recently, injecting tines suitable for grasslands are being used. Injecting is now carried out primarily to reduce smell after spreading, but it can also increase the effectiveness of slurry nitrogen and reduce contamination of grass.

Treatment of slurry along the lines used for urban sewage has been studied extensively over the past two decades. The consensus of opinion at present is that application to land is the only economically practical method of disposing of animal manures. An intensive pig enterprise with 5.000 pig places would have a waste treatment problem equivalent to a town with about 12,000 people. Conventional treatment on this scale would make animal products very expensive. Veal calf wastes are very dilute, by comparison with other animal slurries and are being treated successfully in Holland. Some pretreatment of slurry prior to spreading on land may be desirable; for example solid/liquid separation may reduce some of the problems involved in pumping high dry matter slurry. To avoid spread of disease, slurry can be treated with lime before spreading on land.

Waste Characteristics

Many of the factors already mentioned as influencing quantities also influence the composition of manures. The diet, perhaps more than any other single factor, has an important bearing on composition. Type and age of the animal, as well as storage conditions also have important effects on composition. The degree of dilution with water will greatly influence composition of manure expressed on a fresh weight basis; however, degree of dilution will have little effect on results expressed on manure dry matter.

Agricultural wastes differ from most municipal and industrial wastes in that they are usually highly concentrated and relatively low volume wastes. Biochemical oxygen demands can vary from about 1000 parts per million for dilute slurry or dirty water to over 50,000 parts per million for silage effluent and concentrated pig slurry. The biochemical oxygen demand is a laboratory estimate of the quantity of oxygen required to oxidize part of the organic matter in waste in a specified period of time, usually five days (Anon, 1965). Here we will deal mainly with the quantities and composition of animal manures, with some details on straw, silage effluent, compost, and other wastes. The animal wastes are primarily undigested plant materials with a high proportion of bacterial and intestinal cells.

Crop Waste

The plant nutrient composition of agricultural wastes usually has some relationship to the nutrient content of plant materials. This is understandable, as most agricultural wastes originate directly or indirectly from plant materials. There can be considerable variation in the nitrogen (N), phosphorus (P), and potassium (K) content of plant material. For example, animal wastes in developed countries now have a higher content of N, P, K, and other nutrients than they did 50 years ago. This is due to the higher soil nutrient status with the resultant increase in plant uptake due mainly to the increased used of chemical fertilizers. The improved utilization of animal feed (for example, higher digestability) also contributes the higher nutrient content of manures. Legumes normally have higher N content than do cereal grains, reflecting their higher protein content. This is partly due to their ability to fix atmospheric nitrogen gas in symbiosis with rhizobium bacteria in nodules on the legume roots.

Agricultural crops normally contain 1.5 to 4% N, 0.2 to 0.4% P, and 1 to 3% K in the plant dry matter. During digestion or decomposition, much of the organic matter is used up: however, most of the plant nutrients are retained. For this reason, the nutrient content of agricultural waste may be four times higher than in the original plant material. In the more extreme situation, where the waste is burned and only the ash remains, the nutrient concentration on a dry matter basis will be at a maximum. It should be noted, however, that some nutrients, such as nitrogen, may be partly lost during burning or decomposition. Inorganic nutrients (phosphorus, for example) are often added to animal rations and will increase levels in the manure.

It is estimated that between 80 and 95% of the inorganic nutrients ingested by animals in their feed is excreted in the faeces and urine. Growing and lactating animals retain a higher proportion of the nutrients from their feed than store or fattening animals. For example, Gisinger (1960) showed that dairy cattle excreted 80%, 80% and 95% of ingested N, P, and K, respectively. Fattening cattle, on the other hand, excreted 94%, 98% and 98% of the N, P, and K, respectively. The proportion of organic matter excreted will depend on digestability, but on the average, organic matter excreted is equivalent to about 40% of organic matter intake. This explains the higher inorganic nutrient content of manures. The proportion of faeces to urine in manure will influence composition. Urine is normally low in phosphorus and high in potassium, whereas about equal parts of nitrogen may be excreted in the faeces and urine of cattle (Tietjen, 1976). Therefore, manure in which a proportion of the urine was allowed to drain away would be relatively low in nitrogen and potassium.

It is evident that a high proportion of the plant nutrients in animal feed are excreted in the manure and are potentially available for recycling to cropland for production of subsequent crops.

Cattle, Pig, and Poultry Manures

In general, the higher the dry matter, the higher the nutrient content. For example, in Table 7.1 the higher nutrient content of the poultry manure is due mainly to the higher dry matter content. There is usually about the same level of nitrogen and potassium in cattle slurry, and phosphorus content is usually about 25% of the nitrogen and potassium level. Pig manure usually has about the same level of nitrogen and potassium in cattle slurry, and phosphorus content is usually about 25% of the nitrogen and potassium level. Pig manure usually has about the same level of nitrogen as does cattle manure, although pig manure is usually higher in phosphorus and lowers in potassium than is cattle slurry. The difference in the phosphorus to potassium ratio between cattle and pig manure is principally a reflection of the diet of these animals. The relatively low potassium and high phosphorus in pig slurry reflects mainly a cereal diet, and the high potassium in cattle manure reflects the high potassium content of herbage which often makes up a high proportion of the ruminant diet.

TABLE 7.1: Composition of Cattle, Pig, and Poultry Manures.

		Nutrients (kg per 10 tonnes fresh manure)			
Dry Matter (%)		**N**	**P**	**K**	**Mg**
Cattle manure	4-23	24-65	4-18	20-58	2-6
Pig manure	5-25	16-68	6-21	17-36	3-7
Poultry manure	23-68	96-230	24-120	38-116	12-22

Source: Tunney (1977a)

The nutrient ratio of poultry manure is somewhat similar to pig manure and probably reflects the diets of pigs and poultry. The variation in slurry composition can be greater than for solid manures. The variations in composition of cattle slurry from 33 farms and pig slurry from 25 farms were studied in Ireland (Tunney, 1977a). These results showed a very wide variation. In fact, there was over a ten-fold variation in dry matter between the highest and lowest farms.

Waste Availability

There are various estimates of agricultural waste availability. These may be based on crop-residues, net availability on remote sensing techniques which is used in developed countries. Many researchers/investigators have estimated agricultural waste availability in India (Table 7.2).

TABLE 7.2: Agricultural waste availability in India

Agricultural Waste	Quantity (mt.)	Reference
Crop waste (including industrial waste)	301.6	(Maheshwari, 1984)
Animal waste (including fisheries and marine waste)	944.4	(Maheshwari, 1984)
Crop residues (including industrial waste)	108.0	(Vimal and Tyagi, 1984)
Crop waste (cereals only)	289.7	(FAO, 1985)
Crop residues	407.0	(Bhardwaj and Gaur, 1985)
Animal wastes	2018.0	
Crop residues (including industrial waste)	347.2	(SESA, 1992)
Crop waste residues	287.3	(Mendis, 1988)
Animal waste	633.0	(SESA, 1992)

Estimates of agricultural waste availability suggest that the average value for crop wastes is 350 mt and that of animal wastes it is 650 mt. Hence, around 1000 mt of agricultural wastes are available in the country.

Crop Waste

The quantity of wastes produced daily by an animal is a function of the type of animal, its size, its feed, and the temperature and humidity of the environment (Lehr, 1974). Quantities quoted in the literature often refer to total weight or volume of fresh waste produced, including variable quantities of litter or water.

The nature of the relationship between manure production and liveweight gain of the different farm animals is shown in Table 7.3.

Table 7.3 indicates that on average, daily undiluted fresh manure production is equivalent to between 5 and 8% of the animal liveweight. The daily manure dry matter produced is usually between 0.6 and 1.7% of the animal liveweight. These figures are an oversimplification in that there is considerable variation in manure produced as a percentage of animal liveweight depending on the type and age of the animal and other factors. For example, a dairy cow will produce more manure during lactation than when she is not milking (Tietjen. 1966). The value in Table 7.3. Do, however, indicate a relationship between manure production and weight of animal.

TABLE 7.3: Daily Manure Production. Expressed as percentage of animal live weights.

	Dairy Cow	**Beef Feeder**	**Swine Feeder**	**Poultry Layer**	**Broiler**
Fresh Manure	8.2	6.0	6.5	5.3	7.1
Dry matter	1.0	0.7	0.6	1.3	1.7

Source: Muier and Smith (1975)

Intensive livestock rearing tends toward standardization of diet and type of animal. With this tendency, the quantities and composition of manure should be more uniform. The greatest source of variation will probably be from varying quantities of water or litter mixed with the manure.

Cattle

The slurry production rate most often quoted for cattle is in the region of 40 to 45 kg per day per livestock unit (Tunney, 1977a) a livestock unit (L.U.) is equivalent to a 540 kg dairy cow (Attwood and Heavey, 1964). 500 kg beef cattle produce 40 kg. Of undiluted slurry per animal per day. It has been estimated that about 3 tonnes of dungstead manure or 6 tonnes of farmyard manure would be produced by one livestock unit during a 140 day winter period (Tunney, 1977a).

Pigs

The most often quoted slurry production for meal-fed pigs is 4.5 kg per pig per day on the average over the fattening period (Tunney, 1977a). Sauerlandt and Tietjen (1970) studied the quantities of manure produced by fattening pigs of different weights. At an average weight of 87 kg per pig. 4.35 kg slurry composed of 1.73 kg faeces and 2.62 kg urine was produced. In a study in Scotland (Anon, 1976) pigs at 68 kg mean weight produced 4.5 kg slurry per day with 10% dry matter. One sow and piglets produces about 14.3 kg slurry per day (Christensen, 1976).

Poultry

Loehr (1974) indicated that daily fresh poultry manure represents about 5% of bird liveweight. A 2 kg mature bird would produce about 0.1 kg manure daily with 25% dry matter. Broiler droppings are mixed with litter and normally contain about 75% dry matter when removed from the house.

Silage Effluent

The quantity of silage effluent from ensiled grass and other materials is usually related to the dry matter of the ensiled materials; the lower the dry matter, the greater the quantity of effluent. A tonne of ensiled material with 10 to 20% dry matter will produce 260 to 450 kg of effluent (Harrison, 1973). At 20% dry matter, approximately 100 kg effluent per tonne is produced. At over 25% dry matter, there is generally no effluent. Wilting before ensiling not only improves silage quality, but also has the advantage of increasing dry matter and therefore reducing or preventing effluent disposal problems. Straw yield can vary considerably but is usually in the region of 5 tonnes/ha for high yielding crops.

Spent compost from growing mushrooms, along with other composts, are spread on land and can have a significant fertilizer value. The quantities are small by comparison with other agricultural wastes; however, they can be important in localized areas.

Waste Utilization

Chemical constituents like cellulose, hemicellulose and lignin lead to establish the potential of agricultural wastes as food, feed, fodder, fibre and chemical products of economic importance. The calorific value (CV) indicates the fuel utility of agricultural waste. The CV for various agricultural waste materials varies between 14 to 19 MJ/kg. Higher C content favours calorific value of the waste material. The volatile water in plant residues may be around 80 per cent and fixed carbon is 20 per cent. Volatile matter in plant residues contribute towards oil and tar content. The characteristics of common agricultural wastes available in India are mentioned in Table 7.4.

TABLE 7.4 (a), (b) & (c): Characteristics of common agricultural wastes
(a) Physical and thermal characteristics

Agricultural waste (biomass)	Density (kg/m^3)	EMC at 80%	Lower calorific value (MJ/kg)	Proximate analysis (%)		
				FC	VM	Ash
Pigeonpea stalks	180	20.58	17.60	15.12	82.90	1.98
Bagasse	70	34.86	18.30	15.86	79.20	4.94
Cotton sticks	160	27.05	16.00	15.30	81.40	3.30
Groundnut shell	100	-	18.60	11.67	83.90	4.43
Maize cobs	100	28.00	15.30	15.16	83.01	1.84
Maize stalks	50	38.00	16.10	17.10	79.60	3.40
Rice husk	105	29.40	14.10	12.50	71.00	16.50
Rice straw	30	36.70	13.50	11.10	69.70	19.20
Wheat straw	60	34.00	15.70	17.93	73.60	8.47

TABLE 7.4 (b): Chemical characteristics

Agricultural waste (biomass)	Cellulose (%)	Pentosan (%)	Lignin (%)	Al/BZ extractives (%)
Pigeonpea stalks	41.90	10.30	39.30	2.20
Bagasse	40.00	22.60	14.80	15.90
Cotton sticks	41.90	19.00	27.20	9.30
Groundnut shell	36.50	13.90	31.30	12.20
Maize cobs	36.80	27.90	11.20	14.60
Maize stalks	43.40	19.60	22.80	9.30
Rice husk	44.10	17.80	17.20	3.40
Rice straw	41.40	20.4	12.10	5.60
Wheat straw	39.60	24.10	17.00	7.30

TABLE 7.4 (c): Elemental composition

Agricultural waste (biomass)	Nutrient Content (%)								
	C	H	N	Na	K	P	Ca	Mg	SiO_2
Pigeonpea stalks	53.30	4.70	0.60	0.05	0.57	0.08	0.11	0.41	0.68
Bagasse	48.20	6.10	0.20	0.06	0.51	0.04	0.14	0.36	1.30
Cotton sticks	51.00	4.90	1.00	0.09	0.61	0.08	0.12	0.43	1.33
Groundnut shell	41.10	4.80	1.60	0.05	1.20	0.12	0.10	0.40	2.52
Maize cobs	46.20	4.90	0.60	0.03	0.54	0.07	0.08	0.28	0.90
Maize stalks	41.10	4.20	0.60	0.04	0.42	0.05	0.09	0.45	2.00
Rice husk	37.80	5.00	0.30	0.02	0.30	0.03	0.08	0.17	15.60
Rice straw	36.80	5.00	0.59	0.09	2.50	0.06	0.10	0.53	16.77
Wheat straw	43.80	5.40	0.40	0.06	0.78	0.04	0.10	0.35	7.08

Source: Pathak et al. (1990)

The characteristics of the agricultural waste materials make them important in respect of their use and economic value. Agricultural wastes can be put to following use:

1. Farm development (improvement of soil health, soil fertility, soil physical conditions and plant protection)
2. Source of energy and power
3. Animal feed and fodder
4. Human food
5. Industrial products.

Table 7.4 provides information regarding various agricultural wastes and their uses. Table gives information regarding the individual products from agricultural wastes and by-products.

Ideally, agricultural wastes should be recycled to supply nutrients for subsequent crop growth. But this is not always possible. For example, the food for pigs and poultry is usually transported over long distances. In this situation, the bulky animal wastes cannot be recycled back to the land that produced the crop. The land where the pigs were reared may be inadequate or unsuitable for spreading the manure produced. Manure can then become a disposal problem and be a potential source of pollution.

With cattle, the situation is generally better. The land on which the feed is produced is normally available for spreading manure.

Forage crops make up a high proportion of cattle feed, which is usually fed on or near the farm where it is produced. There are, of course, some exceptions to this for example, the feed lots in the United States where feed for fattening animals is transported long distances and the resulting manure is disposed of rather than utilized. On pasture, grazing animals recycle the waste as it is produced. With winter feed for cattle, there is often a simple recycling relationship where the manure can be spread on land to supply the nutrients for the next crop. Manure from cattle and other ruminants is the greatest single source of agricultural waste for most countries. Poultry and pig manure is usually small by comparison. For this reason, most of the agricultural wastes on a world scale can be recycled to the land from which the crop was produced.

Time of Application

To get a good response from waste application, good management practices should be followed. No more than 50 tonnes/ha should be applied in any one application, and there should be at least 30 days between application. Lower application rates would be advised for wet soil conditions or when applying in the winter time on grassland. Heavy rates of wastes, particularly slurry, can block soil pores. This can lead to anaerobic conditions and reduced infiltration capacity, leading in turn, to increased surface runoff during rainfall. The blocked soil pores will usually be cleared within 30 days by micro-organisms breaking down the organic matter. However, if waste is applied at too frequent intervals, soil micro-organisms will not be able to clear the blocked soil pores.

On grassland, wastes should be applied when the grass is short, either after cutting or after close grazing. Waste application can knock down grass when it is at an advanced stage of growth. Recovery may be slow and the yield may be reduced as a result. Furthermore, the grass is likely to remain contaminated with the waste for a longer period of time. Silage should not be cut for at least six weeks after waste application to reduce the risk of contamination and spoilage of the silage. Slurry spreaders distribute slurry more finely, and much of the slurry adhere to the grass surface; farmyard manure spread with muck spreaders falls in small pieces and is not atomized like slurry.

Cattle or pig slurry at rates of 45 tonnes/ha applied six weeks before cutting did not reduce silage quality or animal performance (Tunney, 1976a). In order to reduce surface contamination and disease risks, it is advisable that wastes should be applied at least four weeks before pasture is grazed.

When and where to spread wastes are important practical questions for agriculture. Wastes cannot be spread at some stages of crop growth: it is usually not possible to spread wastes on manure crops near harvesting time. It is also true that some crops are more suitable for waste application than are others.

From a fertilizer point of view, spring application in USA and Europe and *kharif* (rainy season) application in India usually gives best results, supplying maximum benefit from the nitrogen in the waste. However, it is not always possible to spread all waste in the above sections. In pig production, for example, the manure is produced throughout the year, so adequate storage for twelve months would be necessary if all the manure was spread in spring. This would be very costly. In practice, it is customary that manure produced during winter is spread during spring and summer, and that produced in autumn be spread during late autumn. In some situations, where soil conditions are suitable, manure is sometimes spread throughout the year so that very little storage is required.

The pattern of agriculture determines when and where wastes can be spread. Traditionally, most waste was spread on tilled land. With intensive agriculture, in areas like the British Isles- waste is spread on grassland. In countries such as Denmark, where most of the land is tilled, a high proportion of the waste is spread for tillage crops. In India agricultural waste is incorporated in soil for crops.

On grassland, wastes are normally spread on the grass surface. On tilled land, the wastes are either ploughed or incorporated into the soil after spreading. In this way a higher proportion of the nitrogen can be retained for crop production. Spreading wastes on a grass surface can cause problems by reducing palatability of the grass and increasing the danger of spread of disease to grazing animals. Soil conditions will also influence time of spreading.

TABLE 7.5: Agricultural wastes and their uses

Crop	Waste/By product	Uses	Remarks
Plantation industry			
Coconut	Coir dust (50-60% of husk wt.) Shell	Boards, insulation slabs, moulded articles furfural Fuel, fancy articles, charcoal, activated; destructive compound filler for resins and phenolic moulding powder distillation	
	Water (100 nuts yield 157 litres)	Substrate/growth media for culture, rubber, coagulant, potash source in soil, soft drink	
Arecanut	Husk (1 ha 6-7 Qtls)	Hand board, paper insulating board/wool	
	Leafsheath and leaves	Soles and heels of shoes, gin washers in cotton gins	
	Chogaru liquid or kali	Tannin in leather industry	
Cashew nuts	Cashew apple	Juice for beverages	
	Testa (1% of nut)	As tannin in leather industry	
	Shell liquid	Phenolic compounds for use in resins, lacquer, rubber compounding, barks lining, insulating varnishes	
Tea	Tea wastes (3%)	Caffene in drug industry, manure, fuel for burning	
Coffee	Husk or Pulpy mass (70% of fruit)	Manure, cattle feed, human food, moulded articles, tannins, furfural, fuel, oil recovery from spent grounds	

Crop	Waste/By product	Uses	Remarks
Rubber	Rubber seeds (1 t from 4 ha)	Oil, fats for soap and paint	
	Rubber wood (28 cu m per ha)	Fire wood, panels in tea chests, fibre board paper, rayon fibres, ligno-plastic wood	
	Sticks (Dry fibre 5% green plant jute sticks)	Pulp, paper, viscose rayon, hard board, multiply paper board, xylose, resin, wax	
	Leaves	Sterol-glucoside	
	Mill wastes like mill caddies, loom caddies, short root cuttings	Coarse yarn, paper, fabrics	
Field Crops and Trees			
Cotton	Stalks (2.5t- 5 t/ ha)	Fuel, straw board, paper	
	Linters (8% American Cotton) (1.2% Desi Cotton)	High grade cellulose pulp for viscose and acetate rayon, explosives, lacquers, absorbent cotton	
	Cotton leaves	Cattle feed, citric acid, malic acid	
	Flowers	Flavonols (Useful as vitamins)	
	Roots and bark	Gossypol, catechin	
	Stems	Funfural	
	Press wood	Substrate for yeast	
	Cotton dust (20-50 t/25000 spindles/yr)	Compost	
Tobacco waste	Leaf scrap (21% of total)	Manure, insecticidal, Nicotine sulphate	Present use: Agril purposes 12.6%; Destruction 0.5%; Stalks 8%
	Stalks	Paper	

Crop	Waste/By product	Uses	Remarks
	Seeds (125 kg-150 kg per head)	Oil (25% of seed), oil cake, wrinkle finishes, Factice, seed cake as manure	
Rice milling	Rice bran (5-7% of paddy)	Rica bran oil, poultry feed	
	Rice husk (22-25% of paddy)	As fuel, board and paper, soil conditioner, bedding for animals and poultry, animal feed, abrasive, poli shing paddy and metals, carbon, dent, grease remover, insecticide in rice storage, insulation in cold storage, slow cooling steel in-gots, aggregate for light col. uncrete chemical derivatives like activated carbon, sodium silicate, furfural, oxalic acid, substrate for fermentation.	Mainly burnt as fuel in boiler furnaces in rice mills.
	Husk ash	Absorbent of free oil, protective surface oils from insects, cement, insulating bricks, boiler linings, ceramics, silicone, rubber filler material	
Sugar milling	Bagasse (12.17% of sugarcane)	Paper, furfural cattle feed	Fibre/cellulose 65% of d.w. and pith including soluble matter is 35% of bagasse.

Crop	Waste/By product	Uses	Remarks
	Molasses (3% of sugarcane and 1/3rd of qty. of sugar)	Tobacco curing, foundry; acids like acetic, lactic, citric, aconitic, itaconic; butanol, acetone, glycerol, yeast, dextron, monosodium glutamate	It contains 50% sugar, oxalic acid, lactic acid monosodium glutamate, yeast are at present produced in the country.
	Press mud (3% in sulphitation and 7% in carbon-ation mills)	Manure, sugarcane wax, carbon paper from wax.	
Tree wastes	Leaves, branches, tops of filled trees, thinnings, pine needles	Fuel, building boards, chemicals from wood distillates, insulation wool, cushioning material decorative articles	
Log factory wastes	Bark, off cuts, waste veneers, shavings, chips, saw dust, trimming waste cones and lignin		
Minor oil seed bearing perennial trees like sal, neem, mahua, karanzia, kusum, khakan, Nahor, Undi, Kokum, kala, Pisa, Narotics	Oilseeds	Nonedible oil, lipids, purified fat, protein rich seed meal	
Mahua tree	Mahua flowers (0.74 to 1.48 qtls/ tree)	Alcohol, mahua flower cake	
Sisal	Juice and sisal residue after fibre removal	Hecogenin and wax	

Crop	Waste/By product	Uses	Remarks
Lac	Fine dust	Low grade shellac	
	Mud (7 kg of dry sludge from washing 1 qtl. stick lac)	Contains lac dye (8%), lac wax (10%) and lac resin (8-10%)	
Fruit and vegetable industry wastes			
Mango	Green mango as well as ripe mango peels and kernels (50% of wt. of fresh mangoes)	N free extract, starch for mixing in wheat flour and oil (12% wt. of kernel) from kernel. Pectin and syrup from peels.	
Pineapple	Peels, cores and trimmings	Syrup, vinegar, cattle feed	Wastes contain 7-8% sugar
Citrus fruits	Peels, pomace, seeds and cull fruits (40% of production)	Oil and pectin from peels of orange and lime cosmetics, pharmaceutical industry	
Apple	Peels, pomace and cull fruits	Pectin, vinegar and cider	
Aonla	Seeds	Ayurvedic preparations	
Guava	Peels, seeds and pomace	--	
Peach	Peels and cores	Oil	
Plum	Stone	Oil	
Grapes	Stalks, pomace, seeds and rottens	Compost and wine	
Bananas	Peels	Animal feed, syrup	
Green peas	Shells and vines	Animal feed	
Tomato	Seeds and pomace	Compost and seeds	
Onion	Trimmings	Flavourings	
Potato	Peels, rottens from cold storage	Starch alcohol	
Cauliflower and cabbage	Leaves, stalks and trimmings	Cattle feed	
Ladies finger	Stalks	--	

Crop	Waste/By product	Uses	Remarks
Mushrooms	Stalks, cuttings, trimmings	Pickles and soup	
Asparagus	Cuttings	--	
Living animal waste	Animal dung Animal urine	Compost, fuel gas	
	Wool wax (11.1% of raw wax)	Rust preventive, lubricant, wax acids in paints and varnishes, cosmetics, pharmaceuticals, plant growth promoters, detergent for textiles and others	It is a by-product of combing industry
Slaughter house or dead animal wastes	Hides and skin, Meat, Fat	Tannery, leather, meat seal, soap, animal feed, lubricant, household and bakery use, cosmetics	
	Bones & Sinews (15% or body wt.)	Bone meal, glues, gelatine, fertilizer	
	Horns and hooves	Gelatin, buttons, combs, decorations, neatsfoot oil	
	Blood	Blood meal	
Fisheries and marine wastes	Prawn shell and head	Glucosamine hydrochloride (GAH) or rare, sugar, chitosan and shrimp extract, liquid fertilizer, protein from blanch liquour	
	Low grade and misc. fish, lobster wastes, shark lever residues and fish wastes	Fish protein concentrate (FPC), FP Hydrolysate, bacteriological peptone, ensilage for animal feed, fish meal	
Frog wastes	Frog meal, frogoil		
Marine algae		Agar, algin, proteins, weed food, weed meal, weed manure, pharmaceuticals	

Source: NCST (1975)

Time of application of waste is often determined by when soil conditions are dry enough to carry manure spreading equipment. Under some soil and climatic conditions, there may be six months of the year when manure spreading equipment could cause serious damage to the soil surface. Under other conditions, it is possible to spread manure throughout the year without serious soil damage to grass than the same quantity applied in spring or summer. It appears that slurry can deplete the already low oxygen levels in the soil near the grass roots, killing the better yielding grasses, which will be replaced by weeds. It is therefore advisable that if slurry has to be spread on grassland in winter, low rates (say not more than 20 tonnes/ha) should be used.

Rate of Application

The quantities of plant nutrients required for different crop and soil conditions are reasonably well established for most countries. Where soil fertility is adequate, maintenance dressings of phosphorus and potassium are applied annually. Maintenance dressings are rates that will maintain soil fertility over the years and should equal the difference between losses and the nutrient supplying power of the soil. The main nutrient losses are crop removal and loss in drainage water. Nitrogen is more complex because there are many sources of losses and inputs and it is the primary nutrient influencing yield in conditions of high soil fertility. With low soil fertility, extra rates of nutrients, above maintenance dressings, may have to be applied initially for maximum crop yield. Losses of phosphorus and potassium in water from agricultural land are small in comparison with crop and potassium in water from agricultural land are small in comparison with crop uptake and are usually in the region of a few kilograms per hectare per annum (Kolenbrander, 1973). Removal by the crop is related to crop yield and composition. Intensive grass production for winter food could produce dry matter in the region of 15 tonnes/ ha/ annum. This crop could remove over 400 kg. N, 50 kg P, and 400 kg K/ha/annum. Most agricultural soils can supply large quantities of potassium annually from soil minerals. Rates of nitrogen fertilization for agricultural land are normally less than 300 kg N/ha/ annum. Intensive grassland has the highest nitrogen requirement. Cereals usually require in the region of 100 kg N/ha/annum for maximum production. Leguminous crops that fix nitrogen from the air can give good yields without fertilizer nitrogen.

To get maximum benefit from the nutrients in agricultural waste, they should be applied at rates to supply adequate nutrients for crop needs. Rates of application greatly in excess of crop needs are undesirable since that means nutrients are not being used effectively and the risk of soil and water pollution is increased.

TABLE 7.6: Industrial products from agricultural wastes and by-products

SN	Product	Raw Material
1.	Activated carbon	Paddy husk, coconut shell, Bagasse, Groundnut shell
2.	Apple pectin	Peals and Cores from apple sauce canneries
3.	Adhesives, resins, paints and varnishes	Cashewnut shell liquid
4.	Agar-agar	Sea Weeds
5.	Algenic acid	Sea weeds
6.	Animal and poultry feed	Straws, husks, fruit and vegetable wastes, forest wastes, oil cakes, sea products, sugar mill wastes, distillery wastes, meat processing industry wastes, silk worm pupae, deoiled sal meal, broken egg shell, damaged food grains, poultry feather, rice bran
7.	Aplha cellulose	Bagasse
8.	Alcohol	Molasses, plant straws, Mahua etc.
9.	Acetone-butanol	Molasses
10.	Aconitic acid	Molasses
11.	Bagasse concrete	Bagasse
12.	Boards (Both fibre and particle boards)	Bagasse, jute sticks, groundnut shell, saw dust, cereal straw, cereal husk, coconut and arecanut husk, cashew testa, linseed stalk and cotton stalk.
13.	Building bricks	Paddy husk
14.	Beverages	Cashew, apple, coconut water
15.	Bitter-almond oil	Waste from canning of peaches, apricots and cherrys
16.	Baker's yeast	Molasses

SN	Product	Raw Material
17.	Bone glue, dicalcium phosphate, sterile house meal	Bones
18.	Briquettes	Saw dust, paddy husk, bagasse, groundnut shell
19.	Citric acid	Molasses, peels and cores of pineapple, waste lemons.
20.	Caffeine	Tea waste
21.	Charcoal briquettes	Bagasse, paddy husk, groundnut shell, other organic wastes like fallen leaves, twigs etc.
22.	Cystin	Animal hair
23.	Chitosan	Prawn waste
24.	Cushioning material	Pine needles, coconut husk
25.	Charcoal & wood tar etc.	Wood wastes and organic wastes like plant stalks, twigs and fallen leaves etc.
26.	Dextran	Molasses
27.	Decorative and Fancy articles	Saw mill off cuts, coconut shells. cereal straw, ear heads, maize cobs
28.	Electricity	Bagasse, paddy husk, groundnut shell, saw dust, wood, maize cobs, pigeonpea, stalks etc.
29.	Ethyl alcohol	Molasses, cereal straws
30.	Expansion joint filler	Coir dust
31.	Fertilizer	Press mud, rice husk, bran
32.	Fibre board	Bagasse, paddy husk, arecanut husk
33.	Furfural	Wheat straw, sorghum stalk, maize cobs, coconut shells, groundnut shell, cashew testa, jute stick, bagasse, sawdust, paddy husk, cotton seed husk, sunflower seed hulls, oat husk
34.	Fermentation by- products	Molasses, mahua flowers, tamarind leaves
35.	Fish meal	Fish waste
36.	Fuel gas	Animal dung, community waste, maize cobs, paddy husk, wood waste
37.	Glycerin	Rice bran
38.	Household detergent	Paddy husk
39.	Hecogenin	Sisal waste

SN	Product	Raw Material
40.	Insulation	Arecanut husk, defibrated teak bark, pine wood, pine needles, paddy husk
41.	Leather boards	Leather wastes
42.	Lipids associates	Non edible oilseeds of tree origin
43.	Lactic acid	Molasses
44.	Moulded articles	Coir dust
45.	Macaroon paste	Waste from canning of peaches, apricot and cherries
46.	Monosodium glutamate	Molasses
47.	Methane gas	Cow dung, bagasse, paddy straw
48.	Methyl alcohol	Wood distillate
49.	Nicotine sulphate	Tobacco waste
50.	Oxalic acid	Sawdust, pineapple waste, molasses, mahua flower, maize cobs, cannery waste
51.	Oils and fats	Non edible oil seeds of tree origin; neem, mahua, keranjia, kusum, khukan, nahor, undi, kokum, kamla, fisa, rice bran, tobacco seed, rubber seed, animal fat
52.	Organic manure	(a) Animal wastes and by-products: Cattle shed waste (dung, litter and urine), digested slurry, poultry litter, sheep and goat droppings, by products of slaughter houses and animal carcases (blood, waste meal, bones, horns and hooves, skin and hair wastes) (b) Plant and tree wastes: Crop wastes, farm weeds, aquatic weeds, forest litter (c) Agro-industrial wastes/by products: Oil cakes, fruits and vegetable wastes, tea wastes, tobacco wastes, cotton wool, and silk wastes, husk and bran, by-products of sugar industry, distillery effluents, tannery refuse, and saw mill waste (d) Fisheries wastes and sea seeds
53.	Pulp and paper	Bagasse, jute, cotton and pigeonpea sticks; tobacco, maize and sorghum stalk; castor, linseed and banana stem; groundnut shell; sisal fibre, rubber wood; tendu leaves, cotton linters, rice husk

SN	Product	Raw Material
54.	Protein and protein foods from cellulosic wastes	Straw, cobs, grass, coconut fibre, bagasse
55.	Paper board	Bagasse, rice husk, wheat straw
56.	Particle board	Bagasse rice husk, wheat straw
57.	Plastics	Bagasse
58.	Poultry litter and mulch	Bagasse, rice husk
59.	Power alcohol	Molasses
60.	Pectin and cider	Apple pomace, pineapple wastes
61.	Phospatides	Rice bran
62.	Rectified spirits	Molasses
63.	Rum	Molasses
64.	Surgical suture	Mammalian intestines
65.	Sodium alginate	Sea weeds
66.	Silicon tetrachloride	Paddy husk
67.	Silicon flour	Paddy husk
68.	Starch	Banana pseudo stems, mango kernels, maize cobs
69.	Soil amendments	Bagasse, rice husk
70.	Tanimus	Cashew testa
71.	Vinegar and acetic acid	Molasses, peels and cores from canneries apple sauce canneries, pineapple
72.	Titanium tetrachloride	Paddy husk
73.	Wax	Scour liquor, pressmud, lac effluents, rice bran
74.	Yeast	Molasses

In the past, when manures were valued for their plant nutrients, the normal rate of application was between 25 and 50 tonnes/ha/annum or 5 to 10 tonnes of manure dry matter. Now manures are often applied in much higher quantities of disposal rates of up to ten times the traditional rates. However, such rates could not be recommended either from an agricultural or environmental viewpoint. Approximately 100 tonnes of 4% dry matter slurry would be required to supply the same quantity of nutrients as 50 tonnes 8% dry matter slurry. It is clear that the potassium level at this rate is over double the maintenance requirement of most agricultural crops, while the nitrogen and phosphorus rates are high but not excessive. In fact, if only 50% of the nitrogen is available, supplementary fertilizer would be necessary for maximum yield of many crops. If 100 tonnes/ha of pig slurry were used, then 300 kg N, 90 kg P, and 150 kg K would be applied. In this case, the phosphorus level is in excess of the needs of most agricultural crops but the nitrogen and potassium levels are not excessive. The 100 tonnes of cattle slurry would supply 8 tonnes dry matter and the pig slurry would supply 4 tonnes dry matter per hectare based on dry matter.

From the above values, it is evident that the amounts of nitrogen, phosphorus, and potassium in manure are not in the correct proportion for many crops. There is generally too much potassium in cattle slurry and too much phosphorus in pig slurry relative to the other nutrients. To maximize the value of nutrients in manure, cattle slurry should be applied at rates to give adequate potassium for crop need and he supplemented with fertilizer nitrogen and phosphorus. Pig slurry, on the other hand, should be applied at a rate to supply adequate phosphorus and be supplemented with fertilizer nitrogen and potassium. The same approach can be used to estimate the application rate of other wastes so as to maximize the fertilizer benefit of the nutrients present.

Where more wastes are available than are required for the crop nutrient needs, most crops will tolerate higher application rates. For example, cattle slurry could be applied at a rate to supply adequate phosphorus for the crop, and the maximum yield when relying on nitrogen from cattle slurry alone, as high rates of application may depress yield.

A dressing of 40 to 50 tonnes/ha of good slurry is suitable for most root crops. As for silage, cattle slurry normally needs to be supplemented with phosphorus and pig slurry with potassium fertilizer. Nitrogen fertilizer may also be necessary, and the rate will depend on crop requirements. Care is needed when using waste for cereals because of their sensitivity to excess nitrogen and the difficulty of applying accurate rates of nitrogen due to the variation in waste composition.

It is an established fact that organic waste, in addition to supplying essential nutrients, can have beneficial effects on soil physical properties (Azevedo and Stout, 1974). Under most soil conditions, the physical benefits are probably secondary to those of the nutrients present. However, under certain soil conditions, manures can improve soil structure, with the results of an improved soil environment for root growth and nutrient uptake (Flaig *et al.*, 1978). The value of organic wastes for improving soil structure and water holding capacity is often cited. When attempting to evaluate the fertilizer value of wastes in money terms, it is necessary to base the value on the available nutrients in the manure, and it is, of course, more difficult to put a value on physical benefit that may be of value in some soils and not in others. It is sometimes stated that organic wastes have a beneficial effect on soil by building up soil organic matter. However, most scientific evidence indicates that normal levels of organic waste application have minimal effects on soil organic matter content.

Effect on Soil and Crop

Wastes can have considerable effects on the composition of soils and crops. Soil analyses on experimental plots indicated that land receiving cattle slurry had high soil potassium levels and land receiving pig slurry had high levels of phosphorus (Tunney, 1976a). The manganese content of grass from cattle slurry treated plots was significantly lower than with fertilizer treatment (Tunney, 1976b).

Copper sulphate is added to the diet of fattening pigs and the slurry may contain several hundred parts per million on a dry matter basis. Crops receiving pig slurry can contain increased levels of copper. There can also be a higher total removal of phosphorus from pig slurry than from fertilizer or cattle manure treatments. The higher phosphorus removal is probably due to the higher soil phosphorus level on pig slurry treatments.

In addition, the slurry provides a micro-environment that can facilitate seedling growth in otherwise inhospitable conditions. It is important to ensure that the ammonia content of the slurry is not so high that it can inhibit seed germination. Wastes can also be used to provide a micro-environment for seedling development in reclamation of land; it can be particularly valuable in the rehabilitation of mine heaps and other industrially disturbed soils. In these situations, the wastes are often superior to inorganic fertilizers as they have the dual benefits of improving the physical environment of the plant root and also supplying essential nutrients.

Pollution Hazards

The pollution risks of agricultural wastes and wastes in general have received considerable attention in recent years. The effects of nutrient enrichment from wastes on eutrophication of lakes has been given widespread attention. Phosphorus level is usually the limiting factor for aquatic plant growth and nitrogen is probably the most important nutrient, after phosphorus, contributing to eutrophication.

In addition to nutrient enrichment, wastes can cause water pollution by depleting the oxygen in water. Animal manures have a high biochemical oxygen demand (BOD), and relatively small quantities can deoxygenate large quantites of water. Fish kills attributable to wastes usually result from oxygen depletion of water. If the ammonia in wastes enters the water, it makes fish life more susceptible to low oxygen levels.

There are risks of high nitrate levels in ground water if excessive levels of waste are applied to land, the risk of nitrate toxicity in animals is greater on land receiving high rates of animal manures (Hartmans, 1975). There is some danger of soil pollution from copper and other heavy metals.

The odour and general nuisance problems of agricultural wastes may present many problems (Azevedo and Stout, 1974). These problems can be particularly acute where intensive livestock rearing is practiced in areas of high population density. There are chemicals on the market for controlling smells from manure but they are not very effective. Injecting slurry into the soil is perhaps the most effective way of reducing smell.

Slurry gases. Particularly hydrogen sulphide, are toxic and many fatal accidents involving both man and animal have been recorded. The danger is greatest when slurry is being agitated in enclosed buildings. Flies and dust are problems often associated with animal wastes. The dust problem is greatest in arid and semi-arid regions.

The disease hazards associated with animal wastes are well documented (Strauch, 1976). Spreading manures on to pasture land poses definite risks of spreading animal diseases. Microbial diseases and pathogenic worms can be spread through animal manures.

Dead earth worms are sometimes seen on the soil surface after spreading slurry. However, only a small proportion of earth worms are normally killed, and agricultural wastes usually increase the earthworm population of a soil increasing their food supply.

Despite the potential pollution risks, there is considerable evidence to suggest that if livestock wastes are properly managed, alongwith the rates to supply the nutrient requirements of crops, these wastes can be a valuable asset. On the other hand, if managed carelessly and spread at heavy rates, without regard for crop nutrient needs, they can be responsible for serious environmental pollution.

CHAPTER 8

Biogas Technology for Organic Farming

Biogas technology is anaerobic digestion of organic material producing clean fuel for cooking, lighting and running machinery and the bio-digest slurry besides providing plant nutrients both macro and micro to the crops, benefits the soil by bringing about physical, chemical and biological enrichment. This valuable organic manure not only releases plant nutrients slowly but also improves many of the soil properties like structure, water holding capacity, cation exchange capacity, resistance to soil erosion, and also provides energy for the activity of soil microflora, which enrich the soil and in turn increase productivity. The manurial value of the biogas slurry is superior to traditional FYM or compost in increasing crop productivity, and it enriches the soil with ammonical nitrogen and available phosphorus. While chemical fertilizers do increase crop yield over a short period, they do not maintain sustained productivity due to various reasons like deterioration of soil structure, deficiency of micronutrients and disintegration of soil microflora including N fixing and P solubilizing organisms. The slurry can be applied as liquid under wet, air dried and sundried conditions for enhancing the yield of agricultural and horticultural crops and is also found to be suitable as a seed coating media, fortified with Rhizobium, Azospirillum, Phosphobacteria and micro nutrients, which enhances seed germination, seedling establishment and induces vigour to crops. Biogas slurry enriched with inorganic fertilizers and biofertilizers such as a Rhizobium revealed higher crop productivity and sustained soil fertility.

COMPOSITION OF BIOGAS SLURRY

Cattle dung has been used practice in India since long in the form of dung cake as a domestic fuel and as compost manure. With the advancement of science, a technique of anaerobic digestion came into existence in late thirties, through which an efficient gaseous fuel *i.e.* biogas and digested slurry with higher manurial value can be obtained from the digestion of cattle dung. Besides, being rich in major plant nutrients *i.e.* N (1.5 per cent-2 per cent), P_2O_5 (1 per cent), K_2O (1 per cent) biogas slurry is also a good source of micro nutrients like Zn, Fe, Mn and Cu. One tonne of biogas slurry on dry weight basis will provide approximately the following nutrients and the equivalent cost of chemical fertilizer as given in the table below:

Nutrient	One tonne biogas slurry (dry weight basis)	Equivalent chemical fertilizer	Cost Rs.
N	18 kg	40 kg urea	109.40
P_2O_5	10 kg	63 kg SSP	189.00
K_2O	8 kg	14 kg MOP	66.90
			365.30

In tropics, addition of organic manures to soils is very important from fertility and productivity point of view. Since organic manure is of biological origin, it contains nutrients and improves physical, chemical and biological properties of soil. There is, however, wide variation of nutrient availability in these manures due to differences in animal species, feed consumed and size and health of the animal.

In the biogas plant, nutrient recovery represents an economic return to the farmer using this technology. The liquid effluent from the anaerobic digester is a relatively stable product in comparison to the influent. There is now a broader realization that recycled organic nutrients offer an alternative to reliance on chemical fertilizers. In view of increasing prices of chemical fertilizers and withdrawal of subsidies on them, farmers should be provided with technology for enhancing them efficiency specially that of urea or some alternate source. Biogas slurry is one of the alternatives. The manure obtained from biogas plants has more nitrogen compared to conventional composting. It is reported that fermentation increases the ammonia content by 120 per cent and available phosphorus by 150 per cent. The output of sludge per year per plant (10 cm capacity) is roughly 10 cubic meters. The output of effluent is about 14 cubic meters per year.

Aerobic decomposition of organic materials for manure production results in a loss of 55 percent of organic matter and 30 to 50 per cent of nitrogen. Whereas under anaerobic fermentation in the biogas plant, the loss of organic matter is 25 to 30 per cent and almost complete conservation of nitrogen: NH_4 - N is the main soluble end product in anaerobic unit and 15 to 18 per cent of the total nitrogen is converted to ammonia due to the alkaline nature of the slurry. In our country, the dung rarely contains more than one per cent of total nitrogen on dry weight basis, hence the ultimate nitrogen in slurry will not ordinarily increase 1.25 to 1.30 per cent unless other protein-rich organics are used as feed material in the process. It is directly linked to the nitrogen content initially present in such feed material. Freshly digested slurry has mineralised N to the extent of 21 per cent compared with 16 per cent of N in compost which shows the advantage of anaerobically fermented manure over aerobic decomposition.

AGRONOMIC IMPORTANCE OF BIOGAS SLURRY

Several farmers in Gujarat having biogas plants on farm houses are incorporating the fresh slurry with irrigation water, which is to resulting in higher yields as compared to other modes of its application. This is particularly so with paddy nurseries. The application of such mixed slurry irrigation water resulted in higher yields on transplantation as compared to other methods of fertilization. But such method of disposal is possible only when the gas plant is located on a farm house. Location of the plant plays a significant role in efficient and effective utilization of the biogas slurry.

Field experiments were conducted by Kuppuswamy *et al* (1993) to study the effect of biogas slurry and gypsum enriched biogas slurry on rice-blackgram. Biogas slurry at 10 t/ha enriched with gypsum @ 250 kg/ha gave an additional grain yield of 1.80 t/ha compared to control. Gypsum enriched biogas slurry had an edge over slurry alone. The residual effect of FYM on the succeeding blackgram was comparatively better than that of biogas slurry. In another experiment, gypsum enriched biogas slurry in combination with 75 per cent recommended NPK resulted in higher seed yield under rice-blackgram cropping sequence.

Biogas slurry enriched with gypsum recorded maximum tiller number / hill (7.58) followed by enriched FYM and dried biogas slurry. Wet plain biogas slurry at 10 t/ha recorded increased tiller number by 0.27/hill over FYM alone. Muralikrishna Reddy (1987) had also observed an increase in tiller number through soil application of slurry enriched with gypsum. In respect of number of panicles, the enriched biogas slurry registered the maximum panicle number (412.5/m^2). These treatments resulted in an increase of panicles by 88.44 and 27.72 m^{-2} over control and FYM, respectively. FYM enriched with gypsum and dried biogas slurry at 10 t ha^{-1} and wet biogas slurry at 10 ha^{-1} gave identical panicle number. FYM alone produced more panicles than gypsum alone. Increased panicle number in rice due to application of biogas slurry was earlier reported by several researchers (Sahu and Nandu, 1988; Kuppuswamy *et al*, 1991).

In respect of seed and straw yields, the trend as observed in tiller and panicle number was almost reflected. Wet biogas slurry at 10 t/ha enriched with gypsum at 250 kg/ha proved to be distinctly superior to rest of the treatments. This treatment gave an additional seed yield of 1.80 t/ha compared to control. The response was well marked with FYM enriched with gypsum and dried biogas slurry at 10 t/ha recording increased yields by 21 and 18 per cent, respectively over control. Availability of soil ammonical nitrogen and phosphorus from slurry (Zodhy *et al.*, 1984) might have been responsible for fairly higher yields in slurry treatments. Many workers have reported beneficial effect of biogas slurry on rice yields (Muralikrishna Reddy, 1987; Sahu and Nandu, 1988; Kuppuswamy *et al.*, 1991). In straw yield also, enriched biogas slurry gave the maximum straw yield (15.25 t ha^{-1}), followed by FYM enriched with gypsum (14.69 t ha^{-1}), Wet biogas slurry and FYM did not vary from each other. Increased straw yields with slurry application were also reported by Talukder *et al.*, (1988).

Increased yields with gypsum enriched biogas slurry could be due to the cumulative effect of combining biogas slurry, a good source of organic matter (66.5 per cent) with NPK fertilizer and gypsum containing calcium (21.6 per cent) and sulphur (17.2 per cent).

Sufficient data is not available which compares the fertilizer effect of fermented biogas plant slurry and inorganic fertilizers on plant growth. Jewell *et. al.*, (1976) carried out short green house experiments with maize for getting comparative information. He tested two soils, a mardin soil with a pH around 5.0 and honoeye with pH of 7.7 with each soil, 7 treatments were used:

A control, a fermented cattle manure slurry with total nitrogen added to the pots to obtain concentration of 100, 200 and 300 ppm N and the inorganic pots with ammonium sulphate added to obtain N concentration of 100, 200 and 400 ppm N, phosphorous and potassium were supplemented to 100 ppm to avoid effects other then those caused by nitrogen only about 50 per cent of the nitrogen in the fermented residue was in the ammonia for and the reminder was contained in the non-degraded residue as organic nitrogen. The fermented waste had a superior effect on the response of maize in the early stages. At the termination of this series of experiments of 30 days, growth difference was less obvious. However, at the higher application rates, there was more protein per plant than with higher application of ammonia via chemical fertilizers.

The contribution of nitrogen from various organic manures in crop production, FYM. Compost, biogas slurry and sewage sludge performed equally well on equivalent N basis while poultry manure proved a better source. The residual effect of manure on succeeding crop also, the efficiency or FYM.

As compared to that of fertilizer was noted to be 50 per cent. Due to severe loss of organic matter in sub-tropics the practice of applying heavy dose of manure only once in two years needs to be discouraged and it should be applied in small dose annually.

Rehate (1988) carried out experiments with groundnut which included absolute control with no nitrogen, and all the nitrogen of the recommended dose applied through different sources of organic manure like FYM biogas effluent sludge, PMC and urea.

However, the remaining treatments were the combinations of these sources *i.e.* 50 percent of the recommended dose of nitrogen was added through urea and 50 percent through FYM, and PMC. The results showed that biogas slurry combination with urea favourably influenced the growth, nodulation, dry matter, quality over other treatment and gave significantly higher pod yields over all the treatments. The treatments including slurry plus urea or slurry alone and PMC with urea gave 59.49, 50 and 49.83 percent increase in pod yield of groundnut over absolute control, respectively. However these treatments recorded an increase in pod yield of groundnut over chemical fertilizer treatment to the extent of 37.18, 22.53 and 22.27 percent respectively.

TABLE 8.1: Direct and residual effect of bio-digested slurry on rice and blackgram

Treatments	Rice				Blackgram	
	Tiller number /hill	**Panicle number /m²**	**Seed yield t/ha**	**Straw yield t/ha**	**Pod number/ plant**	**Seed yield kg/ha**
Wet bio-digested slurry @ 10 t/ha	6.83	384	7.46	13.03	9.21	422
Dried bio-digested slurry @ 10 t/ha	7.08	385	7.80	13.69	8.94	393
Wet bio-digested slurry @ 10 t/ha with gypsum 250 kg/ha (1:0.025)	7.58	413	8.41	15.25	9.12	402
Farmyard manure @ 10 t/ha	6.58	341	7.33	12.58	10.22	463
Farmyard manure @ 10 t/ha with gypsum 250 kg/ha (1:0.025)	7.26	385	8.00	14.69	10.11	431
Gypsum 250 kg/ ha	6.33	330	6.78	11.54	7.83	294
Control	6.17	324	6.61	11.03	7.70	292
CD (p=0.05)	0.22	0.06	0.19	0.49	1.24	51

Transfer of Biogas Technology

In India it is estimated that over 16 lakh biogas plants have been installed. Assuming 85 per cent plants in operation at a given time, 27.2 million tonnes of enriched biogas slurry is being produced every year, which is valued at over Rs. 225 crores. Thus in order to make the farmers fully aware of the benefits and utility of biogas slurry, there is a need for launching a mass awareness campaign. Ministry of Non-Conventional Energy Sources (MNES) has launched a countrywide demonstration programme in farmers field, through State Nodal Department, fertilizers, Gujarat State Fertilizer Corporation (GSFC), Indian Farmers Fertilizers Co-operative (IFFCO), Regional Biogas Development and Training Centres and Voluntary Organisations mainly FORRAD, New Delhi. Over 3000 such demonstrations have been conducted so far all over the country.

The reports received from the State Nodal Department of Haryana, Himachal Pradesh, Kerala, Karnataka, Madhya Pradesh, Gujarat, Goa, Punjab, Uttar Pradesh and Tamil Nadu have been studied and analysed for different crops for the period 1984-95 to 1990-91 and the percentage increase in crop yield in demonstration plot (slurry treated) is given at Table.8. 2.

TABLE 8.2: Summary of results of demonstrations conducted by State Nodal Departments/Agencies (1984-85 to 1990-91)

Crop	No. of Demonstration	Per cent increase in crop yield in slurry treated plot over untreated plot
Wheat	93	8-38
Paddy	67	18-36
Gram	11	17-11
Peas	10	15-50
Cotton	2	21-23
Potato	16	5-19
Tomato	18	38-42
Onion	6	10-11
Mustard	10	20-23
Groundnut	24	9-25
Watermelon	5	7-25
Sunflower	13	20-21
Cabbage	7	16-17
Banana	2	7-23

Besides increase in yield of various crops, various observations were made during the field demonstrations. As reported by Tripathi (1993), these observations may be summarised as under:

1. Improvement in quality of produce. For example, the size of vegetables like cabbage, tomato, potato, brinjal, etc. was significantly increased.
2. Reduction in weed number in biogas slurry plots,
3. Biogas slurry does not attract insect breeding.
4. Reduces the usage of chemical fertilizers thus indirectly helps in reduction in unit cost of produce.
5. Improves the physical and chemical properties of soil.
6. Biogas slurry has high water holding capacity which helps rain-fed crops.
7. Seeds treated with biogas slurry have better germination rate.
8. Contains larger amount of organic matter and free ammonia than available in composted manure.
9. Biogas slurry is ready in shortest possible time.
10. If night soil and cattle urine is added, N and P_2O_5 availability in biogas slurry is enhanced.

TABLE 8.3: Summary of results of demonstrations conducted by FORRAD (1989-90 and 1990-91)

Crop	No. of Demonstrations	per cent increase in crop yield in slurry treated plot over untreated plot
Wheat	103	25
paddy	79	32
Peas	14	60
Potato	3	43
Tomato	1	71
Groundnut	7	25
Ladies Finger	1	67
Chilies	1	49
Urad	1	1
Bajra	3	37
Maize	10	49

Crop	No. of Demonstrations	per cent increase in crop yield in slurry treated plot over untreated plot
Ragi	8	27
Banana	3	9
Turmeric	1	27
Sugarcane	2	19
Deccanhemp	1	50
Mulbery	1	25
Mustard	2	23
Total	**257**	**35.5**

In Tables 8.2 and 8.3 above, a total number of 541 field demonstrations have shown an overall average increase of 24.2 per cent in the crop yield on a national basis if biogas slurry is utilised. The utilization of biogas slurry reduces the growth of weeds to a great extent as the seeds of weed are digested in the biogas plant which in turn gives a better crop yield. As far as possible wet slurry should be utilized in the field to avoid loss of ammonia.

In the context of reduction of subsidies on chemical fertilizers, biologically supplied or recycled organic nutrients such as biogas slurry offers an alternative to complete reliance on chemical fertilizers only. The manure obtained from biogas plants has higher nitrogen compared to conventional composting. The bio gas plant will prove more economical to the farmer only when both fuel and manure are used effectively and efficiently by him. Delinking these two aspects will make it uneconomical. It is well known that addition of organic manures to the soil improves its physical, chemical and biological properties. However, more research needs to be done to clearly indicate the benefits of the fermented animal waste on crop production.

According to Santosh *et al.* (1993) for making the programme of organic farming/non-chemical farming more meangful, effective and sustainable in terms of productivity also, following should be given due attention:

1. Establishing proper demonstration site in different agroclimatic conditions.
2. Developing efficient and suitable methods of slurry

management and planning for its popularisation.

3. Conservation of dung resource through the use of improved cooking devices.
4. Popularisation of improved methods of composting such as 'Nadep' technology alongwith biogas.
5. Documentation and scientific assessment of traditional farming systems, especially agricultural practices, cropping pattern, pest management, etc.
6. Maintenance unit for biogas plants in different locale.

CHAPTER 9

Waste Water Treatment and Use for Organic Farming

The main objective of waste water treatment is generally to allow human and industrial effluents to be disposed of without danger to human health and to the natural environment. Irrigation with waste water is both disposal and utilization and indeed is an effective form of waste water disposal. But, some treatment must normally be provided to municipal waste water before it can be used for agriculture. The quality of treated effluent used in agriculture influences the operation and performance of the waste water-soil-plant or aquaculture system.

The most appropriate waste water treatment to be applied before effluent use in agriculture is that which will produce an effluent meeting the recommended microbiological and chemical quality guidelines both at low cost and with minimal operational and maintenance requirements (Arar, 1988). Adopting as low a level of treatment as possible is especially desirable in developing countries.

In the case of irrigation, the required quality of effluent will depend on the crop or crops to be irrigated, the soil conditions and the system of effluent distribution adopted. Through crop restriction and selection of irrigation systems which minimize health risk, the degree of pre-application waste water treatment can be reduced. A similar approach is not feasible in aquaculture systems and more reliance will have to be placed on control through waste water treatment.

WASTE WATER TREATMENT

Conventional waste water treatment consists of a combination of physical, chemical and biological processes and operations to remove solids, organic matter and sometimes nutrients from waste water. General terms used to describe different degrees of treatment, in order of increasing treatment level, are preliminary, primary, secondary, and tertiary and/or advanced waste water treatment. In some countries, disinfection to remove pathogens sometimes follows the last treatment step.

Principal Treatment Methods

Untreated waste water is a hazard to both public health and to the environment. The enhanced waste water production stemming from increasing population and industrialization, together with intensified environmental awareness, have resulted in the enforcement of waste water treatment by stringent water pollution laws and regulations in most developed countries.

Waste water treatment processes are conventionally classified into primary, secondary, tertiary and quaternary or advanced treatments (Shuval, 1977). The higher the level of treatment, the higher the quality of the effluent and consequently, its value. However, a more advanced treatment involves higher costs for construction, maintenance, operation and energy consumption, and the costs rise steeply with each advance in treatment level. The cost of tertiary treatment, including nitrification, denitrification, carbon adsorption and disinfection, can be threefold or more than of simple secondary treatment using oxidation ponds, and about eightfold that of anaerobic facultative ponds (Shelef, 1977). Thus, the necessity for advanced treatment should be carefully assessed according to the intended reuse.

Primary treatment removes coarse organic and inorganic solids, grease and oils from waste water by screening, settling and flotation processes, and also conditions waste water through neutralization and/or equalization before either disposal or discharge to a secondary treatment facility.

Secondary treatment involves both aerobic and anaerobic biological processes, in which complex organic matter in the waste water is decomposed or oxidized by simple microorganisms. Conventional secondary treatments utilize activated sludge, trickling filters, aerated lagoons and oxidation ponds.

Tertiary treatment employing chlorination, microscreening or filtration, coagulation, precipitation and activated carbon adsorption, further removes suspended particles, BOD, nutrients, eutrophication factors and turbidity, and virtually eliminates residual pathogens.

Quaternary or advanced treatment aims at upgrading the effluent to the level of fresh potable water and employs techniques such as ultrafiltration, ion exchange, reverse osmosis, electrodialysis or distillation, by which any undesirable constituent, including excess salinity, can be removed from the water.

The level of treatment required depends mainly upon the intended use or the environmental regulations for discharge of the effluent. In some cases, water quality criteria are unnecessarily stringent for agricultural irrigation use and may restrict effluent reclamation projects. The percolation of waste water effluent through the soil profile, especially the active root zone, is a very effective renovation process, since chemical, biochemical and adsorption reactions remove or inactivate most pollutants and nutrients are taken up by plants.

Sewage water intended for irrigation is treated mainly in order to protect the public both from consuming contaminated crops and from direct exposure to the applied effluent. Additional reasons for treatment are the prevention of nuisance conditions, operational problems in the irrigation system and adverse effects on soil and crops.

Criteria for Selection of Treatment Process

In addition to the required effluent quality, selection of waste water treatment processes depends on the following criteria:

1. Waste water characteristics, *e.g.*, BOD, pH, amount of suspended solids and presence of toxic material.
2. Economic factors, such as cost and availability of land (certain biological processes, *e.g.*, stabilization or oxidation ponds, are only economically feasible if low-cost land is available.

3. Specific public requirements, *e.g.*, complete elimination of offensive odours.
4. Considerations of a possible future upgrading of the effluent.

A summary of standards pertaining to renovated waste water for agricultural use in different countries was published by the WHO (1973, Engleberg Report, 1985; Pescod, 1986; Shuval *et al.*, 1986; Shuval, 1987; WHO Scientific Group, 1989).

Primary Treatment

Primary treatment, intended to remove grit and grease, is usually employed as a pre-treatment for biological processes. The raw sewage water passes through bar racks and/or communicators into grit chambers, where sand and other heavy particles are settled out. The effluent is then discharged into primary clarifies or settling basins, or into flotation chambers where it is detained (usually for several hours) to allow the lighter and smaller solids to settle and grease and oils to rise. Grease and oils are skimmed off mechanically and the heavier solids are scraped off the bottom and delivered to a sludge handling and disposal facility.

Primary-treated effluent has a gray milky colour which does not affect its use for irrigation but does indicate the presence of a high BOD which, under poor management, produces offensive odours. Primary effluent still contains 50-75% of the BOD, 50-65% of the COD and 30-50% of the total suspended solids found in raw sewage water (Asano *et al.*, 1985). Although 50-70% of the total suspended solids of raw waste water are removed by primary treatment, concentrations > 100 mg/l are found in some primary-treated effluents (Table 9.1).

Of the heavy metals present in the incoming waste water, 40-50% is removed into the sludge accumulated at the bottom of the settling basins. Primary-treated effluent is acceptable for irrigation in isolated locations with restricted public access and is limited to crops not intended for direct human consumption. Primary treatment may be the only requirement if storage with a detention time longer than 2 weeks and impoundment depth <4 m is considered, since storage reservoirs act as stabilization ponds in which biological processes substantially reduce pathogenic microorganisms, BOD and suspended solids.

Table 9.1: Properties of primary effluents

Concentration (mg/l)		
Constituent	**Range**	**Median**
Solids		
Total dissolved	200-1500	500
Total suspended	50-150	100
BOD	65-200	135
COD	150-750	335
Nitrogen		
– Total	10-60	40
– Free ammonia	7-40	30
Nitrate	-	<0.1
Phosphorus, total	5-17	8

Source: Thomas and Law (1977)

The objective of primary treatment is the removal of settleable organic and inorganic solids sedimentation, and the removal of materials that will float (scum) by skimming. Approximately 25 to 50% of the incoming biochemical oxygen demand (BOD), 50 to 70% of the total suspended solids (SS), and 65% of the oil and grease are removed during primary treatment. Some organic nitrogen, organic phosphorus, and heavy metals associated with solids are also removed during primary sedimentation but colloidal and dissolved constituents are not affected. The effluent from primary sedimentation units is referred to as primary effluent.

In many industrialized countries, primary treatment is the minimum level of pre-application treatment required for waste water irrigation. It may be considered sufficient treatment if the waste water is used to irrigate crops that are not consumed by humans or to irrigate orchards, vineyards, and some processed food crops. However, to prevent potential nuisance conditions in storage or flow-equalizing reservoirs, some form of secondary treatment is normally required in these countries, even in the case of non-food crop irrigation. It may be possible to use at least a portion of primary effluent for irrigation if off-line storage is provided.

Primary sedimentation tanks or clarifiers may be round or

rectangular basins, typically 3 to 5 m deep, with hydraulic retention time between 2 and 3 hours. Settled solids (primary sludge) are normally removed from the bottom of tanks by sludge rakes that scrape the sludge to a central well from which it is pumped to sludge processing units. Scum is swept across the tank surface by water jets or mechanical means from which it is also pumped to sludge processing units.

In large sewage treatment plants (> 7600 m=3/d in the US), primary sludge is most commonly processed biologically by anaerobic digestion. In the digestion process, anaerobic and facultative bacteria metabolize the organic material in sludge, thereby reducing the volume requiring ultimate disposal, making the sludge stable and improving its dewatering characteristics. Digestion is carried out in covered tanks (anaerobic digesters), typically 7 to 14 m deep. The residence time in a digester may vary from a minimum of about 10 days for high-rate digesters (well-mixed and heated) to 60 days or more in standard-rate digesters. Gas containing about 60 to 65% methane is produced during digestion and can be recovered as an energy source. In small sewage treatment plants, sludge is processed in a variety of ways including- aerobic digestion, storage in sludge, lagoons, direct application to sludge drying beds, in-process storage (as in stabilization ponds), and land application.

Secondary Treatment

Secondary (biological) treatment is the more general requirement in the case of waste water intended for irrigation. The simplest secondary treatment is provided by oxidation ponds, in which oxygen is produced by natural surface aeration and by algae through photosynthesis, involving consumption of nutrients from the waste water. The oxygen is used by bacteria for decomposition of organic matter and the nutrients released are consumed by the algae, thus forming a symbiotic cycle. This type of treatment is suitable for rural communities and small or medium-size urban municipalities (upto about 40,000 inhabitants), in regions of hot to moderate climate and no land shortage.

The objective of secondary treatment is the further treatment of the effluent from primary treatment to remove the residual organics and suspended solids. In most cases, secondary treatment follows primary treatment and involves the removal of biodegradable dissolved and colloidal organic matter using aerobic biological treatment processes. Aerobic biological treatment is performed in the presence of oxygen by aerobic microorganisms (principally bacteria) that metabolize the organic matter in the waste water, thereby producing more microorganisms and inorganic end-products (principally CO_2, NH_3 and H_2O). Several aerobic biological processes are used for secondary treatment differing primarily in the manner in which oxygen is supplied to the microorganisms and in the rate at which organisms metabolize the organic matter.

The oxidation pond system is widely used in Israel. Raw sewage water is conveyed into small anaerobic primary ponds, where grit and sludge settle at the bottom. The anaerobic ponds, usually constructed in a parallel pair to enable alternate removal of sludge, are loaded with 500-2500 kg BOD_5 ha^{-1} (the common load in Israel is 1500-2000 kg BOD_5 ha^{-1} day^{-1}). The depth of the anaerobic pond ranges from 1.5-4.0 m (in Israel, a depth of 2.5 m and detention time of 1 day are usual). The rate of anaerobic biodegradation increases very steeply with increasing temperatures in the 14-40°C range, while at temperatures below 12°C the process virtually ceases.

From the anaerobic ponds, the primary effluent, with a 50-60% reduced BOD_5, flows into a successive series of facultative-oxidation ponds, in which upper aerobic zones are formed above the deeper anaerobic layers. The degree of aerobiosis (depth of aerobic layer), which generally increases in the successive ponds, depends on oxygenation and this in turn depends on solar radiation. The last pond in the series may become completely aerobic and is then referred to as a polishing or maturation pond.

When these treatment systems are well operated, they produce very little, if any, offensive odour. However, for complete elimination of such odours, the anaerobic pond can be omitted by limiting the organic load on the first pond to not more than 180 kg BOD_5 ha^{-1} day^{-1}.

One of the most common secondary treatments employed in larger communities is the activated sludge process, a continuous operation based on recycling of the biological sludge. The so-called extended activated sludge process uses a much larger reactor than is conventionally used, so that both the detention time and aeration are increased; this minimizes the amount of excess sludge, since all degradable sludge formed is consumed by endogenous respiration. The purification capacity of this method is high, but it needs a careful operational control. The area required for such systems is small (about 50 $m^2(10^3 m^3\ day^{-1})^{-1}$).

Another common secondary treatment, suitable for low-flow-rate waste water, uses trickling filters, which are packed with coarse media covered with biological slime, through which waste water is percolated. The 0.1-2 mm-thick slime layer consists of aerobic and anaerobic sublayers. A combination of trickling filters followed by activated sludge is found advantageous for large cities.

Properly treated secondary effluent has very little odour, is clear and almost colourless. Clarified secondary effluent usually contains less than 30 mg/l BOD_5 and 20-30 mg/l suspended solids. Although more than 99.9% of pathogenic microorganisms found in raw sewage water are removed during treatment, the remaining coliform concentration is still very high and may exceed 10^5/100 ml.

Disinfection is an effective means to further reduce bacteria and viruses in secondary effluent. Chlorine is the most commonly used disinfectant and, with sufficient amounts and contact time, it is possible to meet bacteriological discharge requirements and greatly reduce health risks. Applying 10-15 mg/l free chlorine for 5 min at pH 6.5 eliminates 99% of the spores of fungal plant pathogens.

Disinfection normally involves the injection of a chlorine solution at the head end of a chlorine contact basin. The chlorine dosage depends upon the strength of the waste water and other factors. Ozone and ultra violet (UV) irradiation can also be used for disinfection but these methods of disinfection are not in common use. Chlorine contact basins are usually rectangular channels, with baffles to prevent short-circuiting, designed to provide a contact time of about 30 minutes. However, to meet advanced waste water treatment requirements, a chlorine contact time of as long as 120 minutes is sometimes required for specific irrigation use of reclaimed waste water. The bactericidal effects of chlorine and other disinfectants are dependent upon pH, contact time, organic content, and effluent temperature pathogens. In Israel, secondary effluent from oxidation ponds is treated with 20 mg/l chlorine for about 60 min, which virtually removes bacteria and viruses (Kott, 1970) and enables such effluent to be used for irrigation of crops for direct human consumption. No deleterious effect on soil biological activity or crop growth due to irrigation with chlorinated effluent has been reported. Yet, residual chlorine may involve leaf-tip burn and damage to sensitive crops if the free available chlorine (Cl_2) exceeds 0.05 mg/l. Furthermore, toxic effects of chlorinated organics in relation to ground water pollution are also considered (Petygrove and Asano, 1985). Ozone (O_3), produced at relevant treatment plants by passing dry air between two high-potential electrodes, is also used for disinfection. This method is popular in Europe. Its main advantage is the absence of hazardous pollutants.

Tertiary Treatment

Tertiary and/or advanced waste water treatment is employed when specific waste water constituents which cannot be removed by secondary treatment must be removed. Individual treatment processes are necessary to remove nitrogen, phosphorus, additional suspended solids, refractory organics, and heavy metals and dissolved solids. Because advanced treatment usually follows high-rate secondary treatment, it is sometimes referred to as tertiary treatment. However, advanced treatment processes are sometimes combined with primary or secondary treatment (e.g., chemical addition to primary clarifiers or aeration basins to remove phosphorus) or used in place of secondary treatment (e.g., overland flow treatment of primary effluent).

Tertiary treatment is not necessary for **reuse of treated sewage effluent for agricultural irrigation. However, if required**, further removal of suspended solids, BOD and nutrients remaining in the secondary clarified effluent can be achieved by flocculation, which utilizes a coagulant such as alum, polyelectrolyte or lime, followed by a gravel and sand filter or any other graded filter to remove the flocs. This treatment is effective for removing bacteria and viruses which collect in the flocs, as well as most of the suspended solids, BOD, nutrients, heavy metals and turbidity. If lime is used, the lime-clarified effluent, with pH of approx. 11.5, must be neutralized with CO_2. In a plant located at the Greater Tel Aviv area in Israel, sewage water designated for ground water recharge is neutralized by dissolving atmospheric CO_2 into the water in the polishing ponds to lower the pH. During the high pH period, ammonium is converted to free ammonia which is slowly released into the atmosphere. Within 2 weeks in summer, almost all NH_4^+ is thus removed, and in winter about 80%. The high pH environment which prevails prior to neutralization kills bacteria and viruses so that chlorination is unnecessary before recharging the ground water, and the potential hazard of halogenated organics are reduced.

Most of the nitrogen in the influent is in the ammonia form, and this passes through the first two zones virtually unaltered. In the third aerobic zone, the sludge age is such that almost complete nitrification takes place, and the ammonia nitrogen is converted to nitrites and then to nitrates. The nitrate-rich mixed liquor is then recycled from the aerobic zone back to the first anoxic zone. Here denitrification occurs, where the recycled nitrates, in the absence of dissolved oxygen, are reduced by facultative bacteria to nitrogen gas, using the influent organic carbon compounds as hydrogen donors.

The nitrogen gas merely escapes to atmosphere. In the second anoxic zone, those nitrates which were not recycled are reduced by the endogenous respiration of bacteria. In the final re-aeration zone, dissolved oxygen levels are again raised to prevent further denitrification, which would impair settling in the secondary clarifiers to which the mixed liquor then flows.

Activated carbon sorption is used to remove persistent organics from treated effluent. Activated carbons are prepared carbonaceous materials such as, wood, lignite, coal and nut shells, by a thermal activation process that yields a very porous material with a large specific surface area (as high as 1000 m^2/g) onto which solute molecules are adsorbed. Due to high production costs, this process is not widely used. However, as the demand for new water resources increases and effluent quality requirements become more stringent, the use of both activated carbon sorption and of more sophisticated and costly quaternary treatment methods will increase.

TABLE 9.2: Quality of effluents resulting from advanced waste water treatment in USA and Israel.

California, USA		Tel-Aviv, Israel	
Parameter*	**California (Range)**	**Tertiary treatment (Range)**	**Tertiary treatment and ground water recharge**
BOD	2-9	3	<1
COD	-	63-85	20
Suspended solids	1-5	10-13	0
NH3-N	0.1-16.6	3-10	0.1
NO_3-N	0.4-21.3	-	-
Total-N	-	6-13	5
Total-P	-	0.2-1.0	0.1
B	0.6-1.33	0.27-0.29	0.25
Alkalinity	150-256	110-122	-
EC (dS/m)	1.01-1.44	-	-
Coliforms MPN/100 ml	2-4	-	-

*Values given in mg/l, except as noted
Source: California (Asano et al., 1985) and Israel (Ideloniteh, 1978)

NATURAL BIOLOGICAL TREATMENT SYSTEMS

Natural low-rate biological treatment systems are available for the treatment of organic waste waters such as municipal sewage and tend to be lower in cost and less sophisticated in operation and maintenance. Although such processes tend to be land intensive by comparison with the conventional high-rate biological processes already described, they are often more effective in removing pathogens and do so reliably and continuously if properly designed and not overloaded. Among the natural biological treatment systems available, stabilization ponds and land treatment have been used widely around the world and a considerable record of experience and design practice has been documented. The nutrient film technique is a fairly recent development of the hydroponic plant growth system with application in the treatment and use of waste water.

Stabilization Ponds

A World Bank Report (Shuval et al., 1986) came out strongly in favour of stabilization ponds as the most suitable waste water treatment system for effluent use in agriculture. Stabilization ponds are the preferred waste water treatment process in developing countries, where land is often available at reasonable opportunity cost and skilled labour is in short supply.

Waste water stabilization pond systems are designed to achieve different forms of treatment in up to three stages in series, depending on the organic strength of the input waste and the effluent quality objectives. For ease of maintenance and flexibility of operation, at least two trains of ponds in parallel are incorporated in any design. Strong waste waters, with BOD_5 concentration in excess of about 300 mg/l, will frequently be introduced into first-stage anaerobic ponds, which achieve a high volumetric rate of removal. Weaker wastes or, where anaerobic ponds are environmentally unacceptable, even stronger wastes (say up to 1000 mg/l BOD_5) may be discharged directly into primary facultative ponds. Effluent from first-stage anaerobic ponds will overflow into secondary facultative ponds which comprise the second-stage of biological treatment. Following primary or secondary facultative ponds, if further pathogen reduction is necessary maturation ponds will be introduced to provide tertiary treatment.

Suspended and colloidal organic materials in the waste water are removed by sedimentation and filtration through surface grass and organic layers. Removal of total nitrogen and ammonia is inversely related to application rate, slope length and soil temperature. Phosphorus and trace elements removal is by sorption on soil clay colloids and precipitation as insoluble complexes of calcium, iron and aluminium. Overland flow systems also remove pathogens from sewage effluent at levels comparable with conventional secondary treatment systems, without chlorination. A monitoring programme should always be incorporated into the design of overland flow projects both for waste water and effluent quality and for application rates.

The application rate for waste waters will depend principally on the type of soil, the quality of waste water effluent and the physical and biochemical activity in the near-surface environment. Rational design procedures, based on the kinetics of BOD removal, have been developed for overland flow systems by Middlebroods *et al.* (1982). Slope lengths from 30-60 m are common in the USA for overland flow systems.

The cover crop is an important component of the overland flow system since it prevents soil erosion, provides nutrient uptake and serves as a fixed-film medium for biological treatment. Crops best suited to overland flow treatment are grasses with a long growing season, high moisture tolerance and extensive root formation. Reed canary grass has very high nutrient uptake capacity and yields good quality hay; other suitable grasses include rye grass and tall fescue.

Macrophyte Treatment

Maturation ponds which incorporate floating, submerged or emergent aquatic plant species are termed macrophyte ponds and these have been used in recent years for upgrading effluents from stabilization ponds. Macrophytes take up large amounts of inorganic nutrients (especially N and P) and heavy metals (such as Cd, Cu, Hg and Zn) as a consequence of the growth requirements and decrease the concentration of algal cells through light shading by the leaf canopy and possibly, adherence to gelatinous biomass which grows on the roots.

Floating macrophyte systems utilizing water hyacinth and receiving primary sewage effluent in Florida have achieved secondary treatment effluent quality with a 6 day hydraulic retention time, water depth of 60 cm and hydraulic loading 1860 m^3/ha d (Reddy and Debusk, 1987). In Europe, the land area considered to be necessary for treatment of preliminary-treated sewage is estimated at 2-5 m^2 per population equivalent to achieve a secondary effluent quality (Cooper *et al.*, 1988).

Floating Aquatic Macrophyte

Floating macrophyte species, with their large root systems, are very efficient at nutrient stripping. Although several genera have been used including Salvinia, Spirodella, Lemna and Eichornia, Eichornia crassipes (water hyacinth) has been studied in much greater detail. In tropical regions, water hyacinth doubles in mass about every 6 days and a macrophyte pond can produce more than 250 kg/ha d (dry weight). Nitrogen and phosphorus reductions up to 80% and 50% have been achieved. In India, studies have indicated that the coontail, Ceratophyllum demersum, a submerged macrophyte, is very efficient at removing ammonia (97%) and phosphorus (96%) from raw sewage and also removes 95% of the BOD5. It has a lower growth rate than Eichornia crassipes, which allows less frequent harvesting.

Emergent Macrophyte

Natural and artificial wetlands and marshes have been recently used to treat raw sewage and partially-treated effluents. Natural wetlands are usually unmanaged, whereas artificial systems are specially designed to maximize performance by providing the optimum conditions for emergent macrophyte growth. The main features of such treatment systems are:

- Rhizomes of the reeds grow vertically and horizontally in the soil or gravel bed, opening up hydraulic pathways.
- Waste water BOD and nitrogen are removed by bacterial activity; aerobic treatment takes place in the rhizosphere, with anoxic and anaerobic treatment taking place in the surrounding soil.
- Oxygen passes from the atmosphere to the rhizosphere via the leaves and stems of the reeds through the hollow rhizomes and out through the roots.
- Suspended solids in the sewage are aerobically composted in the above-ground layer of vegetation formed from dead leaves and stems.
- Nutrients and heavy metals are removed through plant uptake.

The growth rate and pollutant assimilative capacity of emergent macrophytes such as *Phragmites communis* and *Scirpus lacstris* are limited by the culture system, waste water loading rate, plant density, climate and management factors.

Nutrient Film Technique

The nutrient film technique (NFT) is a modification of the hydroponic plant growth system in which plants are grown directly on an impermeable surface to which a thin film of waste water is continuously applied. Root production on the impermeable surface is high and the large surface area traps and accumulates matter. Plant top-growth provides nutrient uptake, shade for protection against algal growth and water removal in the form of transpiration, while the large mass of self-generating root systems and accumulated material serve as living filters.

TABLE 9.3: Growth and nutrient composition of macrophytes

Macrophyte	Biomass		Tissue composition	
	Standing crop	Growth rates	N	P
	t (dw) ha^{-1}	t ha^{-1} yr^{-1}	g kg^{-1}	
Floating macrophytes				
Eichhornia crasspipes (water hyacinth)	20.0-24.0	60-110	10-40	1.4-12.0
Pistia stratiotes (water lettuce)	6.0-10.5	50-80	12-40	1.5-11.5
Hydrocotyle spp. (pennywort)	7.0-11.0	30-60	15-45	2.0-12.5
Alternanthera spp. (alligator weed)	18.0	78	15-35	2.0-9.0
Lemna spp. (duckweed)	1.3	6-26	25	4.0-15.0
Salvinia spp.	2.4-3.2	9-45	-	1.8-9.0
Emergent Macrophytes				
Typha (cattail)	4.3-22.5	8-61	5-24	0.5-4.0
Juncus (rush)	22.0	53	15	2.0
Scirpus (bulrush)	-	-	15	2.0
Phragmites (reed)	6.0-35.0	10-60	18-21	2.0-3.0
Eleocharis (spike rush)	8.8	26	9-18	1.0-3.0
Saururus cernuus (lizards tail)	4.5-22.5	-	15-25	1.0-5.0

Source: Reddy and De Busk (1987)

A study by Jewell *et al.* (1983) proved this to be a viable alternative for sewage treatment. Reed canary grass was used as the main test species and resulted in the production of better than secondary effluent quality at an application rate of 10 cm/d of settled domestic sewage and synthetic waste water. The highest loading rates achieved were equivalent to treating the sewage generated by a population of 10,000 on an area of 2 ha. Plants other than reed canary grass were also tested and those that flourished best in the NFT system were- cattails, bulrush, strawflowers, Japanese millet, roses, Napier grass, marigolds, wheat and phragmites.

Effluent Storage

Although not considered a step in the treatment process, a storage facility is, in most cases, an important link between the waste water treatment plant and the irrigation system. Storage is needed for the following reasons-

1. To equalize daily variations in flow from the treatment plant and to store excess when average waste water flow exceeds irrigation demands; includes winter storage.
2. To meet peak irrigation demands in excess of the average waste water flow.
3. To minimize the effects of disruptions in the operations of the treatment plant and irrigation system. Storage is used to provide insurance against the possibility of unsuitable reclaimed waste water entering the irrigation system and to provide additional time to resolve temporary water quality problems.
4. To provide additional treatment. Oxygen demand, suspended solids, nitrogen and microorganisms are further reduced during storage.

Reliability of Treatment Systems

Waste water reclamation and reuse systems should contain both design and operational requirements necessary to ensure reliability of treatment. Reliability features such as alarm systems, standby power supplies, treatment process duplications, emergency storage or disposal of inadequately treated waste water, monitoring devices, and automatic controllers are important. From a public health standpoint, provisions for adequate and reliable disinfection are the most essential features of the advanced waste water treatment process. Where disinfection is required, several reliability features must be incorporated into the system to ensure uninterrupted chlorine feed.

EFFECT OF IRRIGATION WITH TREATED WASTE WATER ON SOIL, PLANT AND ENVIRONMENT

Salt Accumulation

The increased salt concentration in the soil following irrigation is a direct result of evapotranspiration. All irrigation waters contain salts; however, whereas the water is taken up by plant roots and evaporates, most of the salts remain in the soil. If the water applied is just sufficient to satisfy the crop water requirement, salt will accumulate in the soil until water uptake ceases. Once the crop can no longer absorb water due to low osmotic potential of the soil solution (i.e., high salt concentration) accumulation will virtually cease. Salinity increase and reduced water uptake are interrelated, i.e., a higher salinity level will result in a lower uptake rate,. Whereas a higher uptake rate will result in faster salt accumulation. For a given irrigation water application (Di), enhanced soil salinity will result in an increase in the fraction of the irrigation water moving beyond the root depth (drainage water; Dd). This fraction is called the leaching.

When the saline soil solution becomes diluted following heavy leaching by winter rains, the soil becomes more sensitive to the sodicity effect of rapid particle dispersion and the consequent soil crusting. This process results in runoff and erosion during the winter and if the infiltration capacity is not restored, in difficulties in supplying sufficient water to the soil to satisfy crop needs during the next irrigation season.

Effluent water may contain organic polyelectrolytes which bind divalent cations, increasing the SAR of the water and thus increasing the potential damage to soil hydraulic conductivity.

Solution of the sodic soil problem or prevention of sodicity development is provided by the application of a source of Ca such as $CaSO_4$ (gypsum) or $CaCl_2$, or, in the case of calcareous irrigated soils, an acid or acid-forming materials such as H_2SO_4 (or S), which dissolves the soil lime present in large quantities in these soils. The amount of Ca added to the soil or irrigation water should be sufficient to reduce the SAR to safe levels. In most sewage effluents and SAR pick up of about 2-3 $(mmol/l)^{1/2}$ is partially compensated by an EC pick up, and if the supply water is acceptable for irrigation, no special problems are anticipated with the reclaimed water.

The effect of Na is aggravated by the presence of carbonates and bicarbonates in the irrigation water and by the concentration of CO_2 in the soil gas fraction. Carbonate and bicarbonate tend to precipitate with Ca, especially during the wetting and drying cycles of an irrigation season, thereby increasing the SAR (CO_3 is present only in water of $pH > 9$).

The effect of Na on nutritional imbalance depends not so much on the absolute concentration of Na but on the ratio of Na to Ca plus Mg in the soil solution (SAR). This effect which occurs more in non-woody crop plants is related more to impaired Ca nutrition than to Na excess, is not so crop specific, and is more severe under low rather than under high salinity conditions.

Ion Toxicity

Some crop plants may be sensitive to high concentration of specific ions. The toxic ion causes yield reduction at concentration (osmotic pressure) level which has no effect on yield if a mixture of salts or another salt is present. The Cl, Na and B ions are most prevalent in effluent water which may be specifically toxic, in addition to some minor elements present in low concentration in some effluents.

Chlorine

One of the causes of chloride injury is thought to be nutritional imbalance. Although there is considerable evidence that the increase in Cl– uptake by some crop plants causes reduced NO3– uptake resulting in nitrogen deficiency, it has yet to be demonstrated quantitatively that the effect is significant and can be corrected by nitrogen fertilization. This nutritional imbalance is most likely to be of importance for very sensitive crops, such as a vocado, which are injured by very low levels of chloride.

Sodium

The Na hazard is threefold: (1) sodium may be specifically toxic in the same way that Cl is; (2) it may cause nutritional imbalance through impaired Ca or K nutrition, and (3) it may affect the physical conditions of the soil, resulting in poor aeration and inadequate water supply.

Boron

Boron is an essential nutrient for most crop plants but toxicity appears when its concentration in the irrigation water exceeds 0.5-1.0 mg/l. The range of optimum boron concentration between deficiency and toxicity is thus rather narrow.

It has been shown (Kenen and Bingham 1985; Kenem *et al*., 1985) that the higher the clay content in the soil, the lower is the boron activity in the soil solution for any given amount of B added to the soil. Plant B uptake and the resulting injury is directly related to B activity in the solution. Thus, the higher the clay content, the lower will be the expected B injury for any given amount of B added with the effluent water.

TRANSPORT OF CHEMICALS AND HAZARDOUS MATERIALS IN THE SOIL

Ground water pollution is a major concern under modern intensive agricultural conditions. Water flowing below the root zone of crops carries various contaminants to the vadose zone and eventually further to aquifers. An understanding of the transport of nutrients in the soil is important for a good irrigation-fertilization management, particularly when using sewage effluents.

Nitrogen

Nitrogen is an essential nutrient element for living organisms, it is usually added in large quantities to agricultural lands to obtain high crop yields. Irrigation with secondary sewage effluent results in the addition of considerable amounts of N to the soil, depending on the content of N in the effluent and the volume of water applied. Consequently, the actual quantity added to crops varies widely and is greatly affected by climatic conditions, soil and effluent properties, crop and irrigation management.

Since effluent N eventually enters the N cycle in soils, irrigation with sewage effluent affects nitrogen uptake by the crop, the level of available N in the soil, gaseous N losses and N leaching below the rooting depth. Irrigation management determines the distribution of N application over time, since N is added in every irrigation with sewage effluent. It is clear that appropriate fertilizer management in effluent-irrigated soils is essential to obtain the most benefit from the presence of N in effluents, and to prevent from the application of excessive or deficient amounts of N.

Nitrogen Transformations in Soil

Level and Composition of Soil N. The total N level, as well as its distribution in the soil profile and its chemical composition, varies among soils. In the top layer of mineral soils, it ranges from 0.2-4 g/kg and consists mainly (90%) of organic forms, while most of the remainder appears as clay-bound non exchangeable NH4. Usually only a small fraction of soil N is present in forms available to plants, namely, exchangeable and water-soluble NH4+ and NO3–.

The soil N substances are components of a dynamic soil N cycle, which is an integral part of the N cycle in nature. The quantities of N added to soils through effluent irrigation can be similar to or even exceed those applied by standard fertilization during similar periods of time. For instance, the amounts of N added in effluent irrigation experiments on cotton (*Gossypium hirsutum* L.) in Israel were often greater than 200 kg $ha^{-1}yr^{-1}$, while the standard N-fertilization range for cotton in Israel is 120-180 kg N $ha^{-1}yr^{-1}$. The quantities of N given to Rhodes grass (*Chloris gayana* Kunth) by irrigation with secondary sewage effluent were even greater, being about 500 kg N $ha^{-1}yr^{-1}$, which is similar to the amounts usually added to this perennial grass as fertilizer (Feigin *et al.*, 1978). However, N fertilization is often necessary due to the presence of low N level in the effluent or when only small quantities of effluent are added, such as in supplementary irrigation of a crop.

As soon as the effluent N reaches the soil, it becomes part of the soil N cycle. The ammonium contained in the effluent, as well as that derived from organic N, is usually oxidized to NO_3^- by nitrification.

Ammonia, derived from NH_4^+ under alkaline conditions, is susceptible to volatilization.

Both NH_4^+ and NO_3^- are taken up by crops, and eventually become immobilized as soil organic matter when crop residues are returned to the soil. The same process occurs in the tissue of soil microorganisms or their by-products. Some of the NH_4^+ may be fixed into a non exchangeable form by certain clays, *e.g.*, the elite group. The significance of this process varies greatly among soils and its effect on the recovery of N added in sewage effluent is probably negligible. Losses of NO_3^- by leaching below the rooting depth (eventually to the ground water) and as gas (N_2 or N_2O) by denitrification are often large.

Ammonia Volatilization. The level of ammonia in effluent depends on the initial NH_4^+ content and the pH. When effluent containing NH_3 is used for irrigation, volatilization may take place. In addition, NH_3 may be formed when effluent NH_4^+ reacts with alkaline soils.

Mineralization-Immobilization. Secondary effluent contains organic N, sometimes at high levels. Organic N molecules are susceptible to microbial decomposition into simple inorganic N compounds, such as NH_4^+ and NO_3^-, which are available to plants (Jansson and Persson, 1982). Heterotrophic soil microorganisms, using organic compounds which contain N as a source of energy, are responsible for these mineralization processes.

Typical organic C/organic N ratios of sewage effluents are around 5 or lower. It may consequently be assumed that these materials are easily decomposable. Thus, the effect of irrigation with sewage effluent is similar to that of fertilization, since addition of a solution with a low C/N ratio will result in addition of available N to the soil.

Nitrification. Most of the effluent N consists of reduced N forms. The predominant N compound in sewage effluent is ammonium, which is a product of the breakdown of organic compounds during mineralization. However, since soil ammonium is susceptible to oxidation, NO_3^- is eventually formed and is the main available N form found in most soils after effluent irrigation.

Irrigation with sewage effluent affects the levels of NO_3^- in the soil, subsoil and ground water, and the level varies with effluent quality and quantity, N fertilization, soil, crop and climatic conditions. Irrigation with sewage effluent greatly increased the level of NO_3-N in the top and deeper soil layers.

Denitrificaion. The transformation of oxidized nitrogen into gaseous N compounds, *i.e.*, denitrification, has recently attracted much attention, and its agricultural and environmental implications are now well recognized. The pathway of denitrification, NO_3^- □ $NO_2^- \rightarrow NO \rightarrow N_2O \rightarrow N_2$, whereby nitrate and nitrite are transformed into gaseous products, is responsible for the loss of up to 70% (average 25-30%) of the nitrate found in soils under normal conditions. Denitrification takes place under anoxic conditions where available NO_3, NO_2, NO or N_2O exist in the soil together with microorganisms.

Addition of organic materials supplying available C to soil microorganisms often results in enhanced denitrification. Thus organic matter is added to effluent intended for ground-water recharge in order to reduce the level of NO_3. Very high denitrification rates occur in soils treated with sewage sludge. Factors reducing gas exchange within soil pores and diffusion of O_2 into the soil tend to increase denitrification rates.

The optimum temperature for denitrification varies greatly among soils and ranges between 50 and 70°C; the minimum temperature is 3-10°C and the maximum about 75°C. The optimum pH is 6-8; however, little denitrification has been reported at a pH as low as 3.5.

Nitrate Leaching from the Root Zone

High NO3 levels in drinking water are hazardous, causing methemoglobinemia in infants. The permissible NO3-N level is 10 mg N/1. An undesirable effect of NO3 in surface water is enhanced proliferation of algae (algae bloom), which occurs in the presence of phosphorus and results in deterioration of potable water quality, even at very low N levels. Nitrates are susceptible to leaching from the plant root zone into deep soil layers and eventually into aquifers. Being an anion, NO3 is not adsorbed by the negatively charged soil clay surfaces, and as its solubility in water is high, it is easily transported through the soil profile. In acid soils (pH

< 5.5-6) or in the presence of amorphous minerals of volcanic origin, positively charged materials capable of adsorbing NO3 are prevalent and NO3 leaching declines.

Uptake of N by Crops

Agricultural crops take up large quantities of N, and only rarely can high yields be obtained without the addition of available N, supplied mainly as fertilizer, manure or compost.

For some crops (*e.g.* fodder) the harvested yield includes the whole plant, while in other cases (*e.g.* cotton) a large portion of the crop remains in the field and is reincorporated into the soil. The actual N uptake varies with variety and growth conditions. The rate of N uptake (kg ha^{-1} day^{-1}) changes during the growing season and it is much greater during certain physiological stages of plant life. For example, the uptake of N by cotton plants, treated by different irrigation and fertilizer combinations, in a trickle irrigation experiment with sewage effluent (Feigin *et al.*, 1984) was above 6 kg ha^{-1} in June, but below 3 kg ha^{-1} day^{-1} later in the season. Since the growth rate of annual crops declines greatly towards the end of the plant life cycle, addition of a large quantity of N by means of effluent irrigation too late in the season may result in an undesirable buildup of NO_3 in the soil. For this reason, among others, the timing of N addition is a very important factor in crop management.

The nitrogen level in plant tissue affects carbohydrate metabolism (Black, 1968). A high N content reduces the carbohydrate concentration in the plant and its storage organs. Feigin *et al.* (1978) noted a great reduction in the sucrose level in sugar beets due to the uptake of large quantities of N derived either from sewage effluent or fertilizer. Furthermore, high N levels in the beets cause processing difficulties. The total yield of sugar and the cane quality obtained from a sugarcane field irrigated by sewage effluent in Hawaii (Lau *et al.*, 1975) were lower than those obtained using fresh water irrigation, even though the total yield was 11% greater.

Excessive N affects crop growth and yield. Enhanced vegetative growth as indicated by plant height and a reduction in yield was found in cotton plants to which excessive amounts of N were added through sewage effluent or freshwater amended with fertilizer. The tall plants had thinner shoots and severe lodging occurred, which is also a frequent problem in cereal crops which are heavily N-fertilized.

Delay in ripening of orange fruit has been associated with excessive N levels (Baier and Fraier, 1973). Maynard *et al.* (1976) reviewed the undesirability of high levels of NO_3 in vegetable crops and concluded that the nitrate level in crop tissue increased as the rate of N fertilizer increased, and was higher with NO_3 addition than with NH_4 addition. The tissue NO_3 levels had not reached the range hazardous to adults.

Phosphorus

Phosphorus is an essential nutrient element for all organisms and is added in large quantities to obtain enhanced crop yields. Secondary sewage effluent often has high phosphorus content, making it an important source of P for irrigated soils. Effluent can partially, or even completely, replaces other P sources and prolonged effluent irrigation results in great savings in P fertilizer and has an important economic role.

The amount of P added to the soil through sewage effluent irrigation is usually not excessive. However, high loads of P are sometimes added, especially when the soil is used for disposal purposes. Excessive levels of available P may result in nutrient imbalances, such as Cu, Fe and Zn deficiencies (Ryden and Pratt, 1980), necessitating soil or foliar application of micronutrients.

Application of P to the soil, as fertilizer, manure or in effluent, results in an immediate rise in the level of water-soluble P in the soil. However, due to adsorption and precipitation reactions taking place in the soil, the level of soluble P declines rapidly with time. Phosphorus uptake by plants also reduces the level of soluble P in the soil, but usually at a much slower rate. The actual level of water-soluble P in soil is usually very low, and the movement of P through the soil is restricted. Consequently, soil P compounds are usually considered immobile and P leaching and runoff have been considered negligible. The transport of even small quantities of fertilizer-derived P from agricultural soils into surface waters can induce eutrophication (algae bloom) in surface waters (Taylor and Kilmer, 1980).

Phosphorus Concentration and Forms in Soils

The level of total P in soils varies greatly, lying between 0.2 and 0.8 g P/kg dry soil and 500-2000 kg P/ha in the top 0.2 m layer. The level of P available to plant as determined by various chemical and biological methods (Olsen and Sommers, 1982) comprises only a small fraction of the total P present. The concentration of water-soluble P is 0.03-3 mg/kg dry soil.

Phosphorus Transformations in Soils

Substances containing phosphorus such as fertilizer, organic amendments and sewage effluent, when added to the soil, soon become an integral part of the P cycle in the soil. A simple model for the P cycle in the soil-plant system suggested by Jones et al. (1984) is used for describing the fate of P added by means of sewage effluent. This model consists of pools representing the different P compounds, having specific chemical and physicochemical properties and vectors indicating relationships between the pools given by fertilization. For instance, 500 mm/yr of secondary sewage effluent adds 50 kg P/ha–1yr–1, if the P content in the effluent is 10 mg/l. However, if the quantities of effluent applied are much greater, as in the case of high-rate applications for disposal purposes, the amounts of P added exceed those usually applied through fertilization, and accumulation of available P in soil may cause development of micronutrient deficiencies (Ryden and Pratt, 1980).

Effect of Irrigation with Waste Water on Soil P

P movement through effluent-treated soils has been studied under different conditions. Sawhney (1977) found that the breakthrough of P from a fine sandy loam and silt loam soil columns occurred after about 50 and 100 pore volumes of effluent, respectively, had passed through. Similar results were obtained by Nagpal (1985) in a column experiment using a 60-cm-deep gravelly silt loam soil. Equilibrium P concentration in the leachate was obtained after about 70 pore volumes of effluent had passed through the columns. Moreover, about 60-90% of the added P was retained by the soil even after 2.7 years of leaching, most of it in the surface soil.

Phosphorus Uptake by Crops

The uptake of P by agricultural crops varies greatly; common values are 10-100 kg ha–1hr–1, with most values being 20-50 kg ha–1yr–1. The total P uptake by crops is lower than that of N and K, its concentration in plant tissues usually being one-tenth to one-fifth of that of N and K. The P taken up may be removed from the field in the harvested portion or may be reincorporated into the soil in the residue. Dag et al. (1984) found that the quantities of P taken up by cotton plants (seed or whole plants) during a 5-year study on a clay soil were considerably greater in effluent-treated plots in comparison with freshwater-irrigated plots. This was the result of both higher P concentration in plant tissue and greater production of dry matter. Nitrogen fertilization was found to influence P uptake in the freshwater treatments. The quantity of P added through standard irrigation with sewage effluent (usually 3-10 mm/day) seldom results in the addition of P above the normal levels

Potassium

Potassium Concentration in Soils

Potassium is an essential nutrient element for organisms, and it is taken up by crops in large quantities, similar to, or even greater than those for N. Since the level of available K in the soil is often below that necessary for production of high crop yields, fertilizers and organic materials (e.g. manure) are often added. Furthermore, as K fertilizer is also considered an effective means for obtaining a high-quality product, it is extremely important in high-productive agriculture. The level of K in sewage effluent varies greatly, typical concentrations being 7-20 mg K/l. Irrigation with sewage effluent often does not satisfy plant needs in K, especially when the level of available K in the soil is low. When higher levels of K are present in sewage effluent, it becomes an important source of the element. K overdose is unusual when standard quantities of sewage effluent are used for irrigation.

Potassium Transformations in Soils

Release of K+ from soil minerals is by exchange, diffusion and dissolution processes. The K level in the soil solution depends on the characteristics of the clay minerals, cation exchange reactions and root activity.

Potassium Uptake by Crops

The potassium percentage in plant tissue is relatively high, which indicates a high crop uptake. Feigin et al. (1979) measured the quantities of K taken up by sugar beet, Rhodes grass and cotton grown in effluent-irrigated soils. The amounts of K removed by sugar beet and Rhodes grass were four and two times greater, respectively than those added through the effluent. The uptake by cotton plants was also two to three times greater than applied, but most of it was later incorporated into the soil with the vegetative parts of the crop.

Palazzo (1977) noted a reduction in the level of exchangeable K in soil following irrigation of forage crops at rates of 50, 85 and 250 mm/week with waste water having a concentration of 8.8 mg K/l; this supplied 114, 172 and 306 kg K ha-1 yr-1, respectively. The reduction was partially related to plant K uptake, which was on the average 247, 244 and 346 kg K $ha^{-1}yr^{-1}$, respectively. The level of K in the crop tissue was also reduced; it was 26.7, 23.7 and 27.2 g K/kg dry matter, respectively, in the first growing season, and 21.0, 22.2 and 19.4 g K/kg in the second.

Trace Elements

Trace elements are usually present in low concentrations in natural systems. Certain of these chemicals are especially important in waste-treated soils. Much of the information on trace elements originates from soil-plant systems treated with sewage sludge.

Page (1974) mentioned the following trace elements as important components of sewage sludge: silver (Ag), arsenic (As), boron (B), barium (Ba), cadmium (Cd), cobalt (Co), chromium (Cr), copper (Cu), mercury (Hg), manganese (Mn), molybdenum (Mo), nickel (Ni), lead (Pb), selenium (Se), tin (Sn), vanadium (V) and zinc (Zn). Certain trace elements, including most of those listed by Page, are heavy metals, with a density greater than 5 g cm^{-3}, although by common usage other metals are also included in this list (Tiller, 1989). The following 12 out of the 38 existing heavy metals are used in industry and are therefore important potential pollutants: Cd, Cr, Co, Cu, Fe, Hg, Mn, Mo, Ni, Pb, Sn and Zn (Jones and Jarvis, 1981). Tiller (1989) lists Cd, Pb, Cu, Zn, Ni, Sb and Bi as most important environmental pollutants.

Trace elements participate in many of the biochemical processes that greatly affect plants, animals and humans. The following trace elements are essential for plant development and growth (and are therefore named micronutrients): Fe, Mn, Zn, Cu, Mo and B. Those required by animals are Mn, Cu, Zn, Fe, Mo, Se, I, Co, F, Cr, Ni and V, Na, Cl, Co, V and Ni are also mentioned as being essential for certain plants (Tiller, 1989). On the other hand, other trace elements (Cd, Hg, Pb included) have been shown to be nonessential to plants and animals. Both the biologically required and the other relevant trace elements become toxic to organising when present at concentrations exceeding certain threshold levels of their available forms. The margin between recommended (or acceptable) and toxic concentrations is often quite narrow, as in the case of B. The health hazard posed by excessive levels of trace elements in soils, and consequently in the tissues of plants, animals and humans consuming the plant materials, is therefore of great public concern (Purvis, 1985). Movement of trace elements to the ground water is also considered as a serious threat under these conditions.

The quantities of trace elements added to the soil depend on the origin of the wastes and waste water used. The concentration of trace elements in sewage water varies greatly (Thomas and Law, 1977; Page *et al.*, 1981). High concentrations are found in industrial sewage effluent, especially in raw sewage, while the levels in domestic effluents and in advanced-treated sewage effluent are very low and negligible, respectively.

Some trace elements are hazardous to organisms, while the toxicity of others is low. This is the result of their biochemical effect on living organisms, their physicochemical characteristics, and the forms and actual quantities added to specific soil-plant combinations. Cadmium, Cu, Mo, Ni and Zn are considered more potentially toxic than other elements applied to soils treated with waste or waste water. For various reasons, including their presence at low level in waste water and the availability of suitable agronomic management practices, Mn, Fe, Al, Cr, As, Se, Sb, Pb and Hg are categorized as posing little environmental hazard (Page *et al.*, 1981). Boron is also mentioned as a potential hazard if added in an uncontrolled manner.

TABLE 9.4: Maximum tolerable levels of dietary minerals for domestic livestock in comparison with levels in forages.

Element	Soil-plant Barrier	Level in plant foliage		Maximum levels chronically tolerated[a]			
		Normal	Phytotoxic	Cattle	Sheep	Swine	Chicken
		(mg/kg dry foliage)		(mg/kg dry diet)			
As_{inorg}	Yes	0.01-1	3-10	50	50	50	50
B	Yes	7-75	75	150	(150)	(150)	(150)
Cd[b]	Fails	0.1-1	5-700	0.5	0.5	0.5	0.5
Cr^{3+}	Yes	0.1-1	20	(3000)	(3000)	(3000)	(3000)
Co	Fails	0.01-0.3	25-100	10	10	10	10
Cu	Yes	3-20	25-40	100	25	250	300
F	Yes?	1-5	-	40	60	150	200
Fe^{2+}	Yes	30-300	-	1000	500	3000	1000
Mn	?	15-150	400-2000	1000	1000	400	2000
Mo	Fails	0.1-3.0	100	10	10	20	100
Ni	Yes	0.1-5	50-100	50	(50)	(100)	(300)
Pb[b]	Yes	2-5	-	30	30	30	30
Se	Fails	0.1-2	100	(2)	(2)	2	2
V	Yes?	0.1-1	10	50	50	(10)	10
Zn	Yes	15-150	500-1500	500	300	1000	1000

[a]Based on NRC (1980). Continuous long-term feeding of minerals at the maximum tolerable levels may cause adverse effects. Levels in parenthesis were estimated (by NRC) by extrapolating between animal species.
[b]Maximum levels tolerated were based on Cd or Pb in liver, kidney, and bone in foods for humans rather than simple tolerance by the animals.

Beneficial and Hazardous Effects of Trace Elements on Plants and Animals

The addition of trace elements by means of effluent irrigation can be beneficial under certain circumstances. Thus, the micronutrient deficiencies occurring under various soil-plant conditions, especially in calcareous soils, can be corrected by adding mineral or chelated trace element substances Since some trace elements are required by animals, their addition through plants irrigated with sewage effluent can be considered as another advantage obtained from waste water use (in addition to the contribution of water and plant nutrients). However, the main concern today is the danger o accumulation of excessive levels of trace elements leading to toxicity and hence to health and environmental hazards. Such problems are not restricted to waste use, but also occur in heavily industrialized areas as a result of other contamination factors (Purvis, 1985).

In addition to the information on plant response to trace elements and their accumulation in plant tissues obtained for raw sewage, or sludge amended soils, laboratory studies have been carried out on solutions and simulated effluent rich in trace elements. Accumulation of trace elements in plant tissue may detrimentally affect crop growth, but some trace metals accumulate in plant material without any external signs. Consumption of such materials through the food chain may adversely affect animals and humans.

Much attention has been paid to Cd and Zn due to toxicity hazard and their potential for accumulating in plant tissue. Other elements considered as most hazardous due to their occurrence in large quantities in wastes such as sewage sludge and their toxicity affects are B, Cu, Mo and Ni.

Effects of Specific Trace Elements on Soil, Plant and Food Chain

Arsenic (As). Cases of arsenic pollution have usually been attributed to pesticide application or to the use of industrial products. Most of the information on the potential toxicity of As is obtained from studies carried out in highly contaminated sites. The range and median levels of arsenic in dry digested sludge are 1.1-230 and

10 mg/kg in comparison with <0.005-0.023 mg/l in municipal secondary effluents). Toxicity may occur after application of 20 mg/ha of high-As sludge. In the soil, the element reverts to arsenate which is strongly held by clay; the availability of as is reduced in fine-textured soils.

Arsenic usually accumulates in the plant roots, so that its content in the edible plant parts is generally below the critical level of 3.0 mg/kg. The typical as levels in plant tissue are between 0.1 and 5 mg/kg dry matter, and in soil it is 6 mg/kg. Its level in forage crop is 0.01-1 mg/kg with toxicity occurring at 3-10 mg/kg. The estimated concentration in soil solution at toxic levels, estimated at 30 times its concentration in sea water (Bohn *et al.*, 1985), is 0.1 mg/l. This is also the recommended maximum concentration in irrigation water (Ayers and Westcot, 1985). Toxic effect on crops varies greatly, *e.g.*, 0.05 mg/l is toxic for rice and 12 mg/l for Sudan grass. Toxicity and yield failure precede the production of crops containing hazardous levels in edible parts. Consequently, the soil-plant barrier protects the food chain against As (Chaney, 1989). Arsenic is usually added to poultry diet.

Boron (B). Boron is essential for plants, but is not required by animals. About 0.5 mg/l is usually added in nutrient solutions used to nurture plants in soilless cultures. Boron is associated with salinity problems and its presence in arid soils is widespread. Since its main form in waste water is probably non dissociated boric acid, $B(OH)_3$ (an uncharged molecule) boron moves faster in soil in comparison with other trace elements. It can be leached out, although leaching requires large quantities of water. Its concentration in crop tissue is 5-30 mg/kg.

Cadmium (Cd). Cadmium is considered as a potentially serious environmental and health hazard. This is because of its mobility in the food chain and to its high toxicity to plants, animals and humans. The incidence of the Itai-Itai disease and other health problems are associated with the presence of Cd in the food chain (Purvis, 1985). The Itai-Itai disease occurred under special conditions in Japan where the water used to irrigate rice originated in a Cd-mining area.

The Cd level in soils varies greatly from 0.01-7.0 mg/kg, the typical concentration being 0.06 mg/kg. The typical Cd concentration in the soil solution is estimated as 0.001 mg/l; in secondary municipal effluents it is <0.005-0.15 mg/l. The recommended upper level in livestock drinking water is 0.05 mg/l. In sludge, the Cd content is about 1-3000 mg/kg with a median value of about 10 mg/kg. The typical concentration of Cd in plant tissues is 0.1-1.0 mg/kg.

Bingham *et al.* (1975, 1976) studied the relation between the levels of available Cd in the soil and in plant tissue. The most sensitive crops, of those tested, (*i.e.*, 50% yield reduction as a result of 10-50 mg Cd/kg soil) were: spinach (*Spinacia oleracea* L.) soybean (Glycine max Merr.), sweet corn (*Zea mays* L.), and upland rice (*Oryza sativa* L.).

The Cd level in animal tissue increases as its level in the diet increases. Cadmium levels in kidneys and liver exceed that in muscles. The quantity of sludge allowed for application is based on predicted Cd intake.

Cobalt (Co). Cobalt is required by leguminous plants (<0.2 mg/kg) and by animals. Its concentration in this soil is 1-40 mg/kg, the typical value being 8 mg/kg. The normal range in plant tissue is 0.1-0.3 mg/kg. The estimated level in soil solution under natural conditions is 0.01 mg/l and its recommended upper level for livestock in drinking water is 1.0 mg/l. The higher acceptable level for long-term irrigation is 0.05 mg/l. Cobalt is toxic to tomato plants at 0.1 mg/l in the nutrient solution, and its toxicity level in plant foliage is 25-100 mg/kg dry matter.

Chromium (Cr). Chromium is not generally considered as an essential element, but deficiency in humans and animals has been reported. Its toxicity hazard is low. Its level in soil ranges from 5-3000 mg/kg, with a typical concentration of 100 mg/kg. Its estimated level in soil solution is 0.01 mg/l. The range of normal level of Cr^{3+} in plant tissue is 0.1-1 mg/kg and phytotoxicity is found at 20 mg/kg in plant foliage (Chaney, 1989). Hexavalent Cr is toxic to plants, but its levels in sludge and natural systems are low to very low. In the soil this element reaches a Cr^{3+} state with low solubility. Due to the limited possibility of hexavalent Cr being present in the soil, Cr does not pose a real hazard to the food chain. The level of Cr in sludge varies greatly. Its estimated level in the soil solution is 0.01

mg/l. Most crops take up small quantities of Cr, but some species can take up considerable quantities; Keefer *et al.* (1986) found 2 mg Cr/kg or more in the tops of radish plants, but levels as high as 30 mg/kg have also been reported. A restricted translocation from leaves to grain was detected in maize. Plants can tolerate high levels of Cr in soil.

Copper (Cu). Copper is required by both plants and animals, but is also considered as one of the elements posing a potentially serious hazard when present in excessive levels, as in the case of certain sludge-treated soils. The range of Cu in soils is 2-100 mg/kg with a typical level of 20 mg/kg. In soil solution its level is 0.03-0.3 mg/l. The level of Cu in sewage sludge is 80-17000 mg/kg with a median value of 800 mg/kg.

Manganese (Mn). Manganese is required by plants and about 0.5 mg/l is added in nutrient solutions. The content of Mn in soils is 100-4000 mg/kg, with 850 mg/kg being the typical level, and in soil solution it is 0.1-10 mg/l. The range of Mn in plant tissue is 15-150 mg/kg and phytotoxic levels in plant foliage range between 400-2000 mg/kg dry matter. The recommended upper level for long-term irrigation is 0.2 mg/l (Ayers and Westcot, 1985). In acid soils, Mn may become toxic to certain plant species. Manganese toxicity is closely related to iron deficiency.

Molybdenum (Mo). Molybdenum is required by plants and is added at low concentrations (about 0.03 mg/l) to nutrient solutions. The normal concentration in soils is 0.2-5 mg/kg, a typical value being 2 mg/kg. The normal concentration of molybdenum in plant tissue is 1-100 mg/kg and in forage foliage it is 0.1-3 mg/kg. The recommended content in irrigation water is 0.01 and 0.05 mg/l for long- and short-term irrigation, respectively.

Despite the low levels needed by crops, high concentrations of Mo are not toxic to plants; toxicity in forage crops occurs at a level of 100 mg/kg dry matter. However, Mo may pose a hazard to animals if present in excessive levels in plants. Animal response to Mo intake depends on the species and the age of the individual. Other influencing factors are the concentration of Cu in the food and sulphate level in the diet. Cattle are most susceptible and chicken are least sensitive to Mo toxicity. Molybdenosis in animals, having the same symptoms as Cu deficiency, may develop when forage has Mo levels as low as 5 mg/kg. Since excessive Mo levels in forage crops may lead to serious problems. Molybdenum is more available in alkaline soils. Molybdenum derived from organic residues poses no hazard to the food chain in drained soils having a pH lower than 6.

Nickel (Ni). Nickel is not required by plants but is needed by animals. The level of Ni in soil is 10-10000 mg/kg, with 40 mg/kg being the typical value. The concentration in the soil solution is estimated to be 0.05 mg/l. Nickel is toxic to some plants at levels between 0.5 and 1.0 mg/l. The normal content of Ni in plant material is 1-10 mg/kg dry matter and 0.1 and 5 mg/kg dry foliage. The level of Ni in soils is controlled by the level of Mn or Fe hydrous oxides that reduce the level of soluble Ni in the soil. The level of available Ni is also affected by the presence of organic matter. The range of concentrations of Ni in sludge is wide (2-5300 mg/kg, with a median value of 80 mg/kg). Nickel is toxic to plants at concentrations greater than 50 mg/kg dry matter. Nickel toxicity is acute only in acid soils and it is reduced by liming. The upper acceptable limits of Ni in irrigation water are 0.2 and 2.0 mg/l for long term and short-term use, respectively.

Lead (Pb). Lead is not required by plants and animals, but is potentially highly toxic to animals and less so to plants. The high Pb toxicity is mainly associated with industrial pollution sources rather than with the development of excessive Pb in soils. The level of Pb in soils ranges between 2 and 200 mg/kg, with a typical value of 10 mg/kg, and its concentration in the soil solution is estimated as 0.001 mg/l. The recommended upper level of Pb in animal drinking water is 0.1 mg/l, but problems may occur with levels as low as 0.05 mg/l. The maximum acceptable concentrations in the irrigation water are 5 and 20 mg/l for long- and short-term applications, respectively. The typical median content of Pb in sludge is 250 mg/kg.

Increasing the soil pH, as well as the CEC and P levels, results in a decrease in soil Pb availability. The lead level in soils is reduced by liming. Large applications of Pb to a soil having a pH level of 5.0 does not result in yield reduction or accumulation of lead in plant tissue. The highest level of Pb is found in plant roots, and its value in other plant parts diminishes in the order; shoot, fruit and seeds. Its typical concentration in crop tissue is 0.1-10 mg/kg and 2-5 mg/kg in forage. Under normal conditions of sludge application, when a large quantity of P is also added and at soil pH above 5.5, the movement of Pb to plant tops and seeds is nil.

Selenium (Se). Selenium is not required by plants, but it is essential for animals at low concentrations (0.04 mg/l). It becomes toxic to animals at levels of 4-5 mg/l. The range of Se in soils is 0.1-2.0 mg/kg with 0.5 mg/kg being typical; the Se level in plant tissue is 0.02-2.0 mg/kg with toxic levels above 50 mg/kg. The maximum recommended content in animal drinking water is 0.05 mg/l, and 0.02 and 0.05 mg/l in long-term and short-term irrigation, respectively. The level of Se in soil solution is 0.001-0.01 mg/l and that in digested sludge ranges from about 2 to about 17 mg/kg. The level of Se in waste water is usually very low.

Vanadium (V). Vanadium is required by certain algae. It is not required by plants, but is toxic to many plants at a relatively low concentration. Its level in soils ranges from 20 to 500 mg/kg with a typical concentration of 100 mg/kg. The tolerable concentration in animal drinking water and recommended maximum concentration in long-term irrigation water is 0.1 mg/l. Its levels in the tissue of forage crops are 0.1-1 mg/kg, and it becomes toxic at a concentration of 10 mg/kg.

Zinc (Zn). Zinc is essential for plants and animals, and is added at the rate of about 0.5 mg/l to nutrient solutions. Its concentration in soils is 10-300 mg/kg, with 50 mg/kg being the typical level. Zinc concentration in soil solution is <0.005 mg/l. The typical range of Zn in plant tissues is 15-200 mg/kg. The recommended maximum value for drinking water for livestock is 24 mg/kg. The recommended maximum value for drinking water for livestock is 24 mg/l, and for water used for long-term and short-term irrigation 2 and 10 mg/l, respectively. Zinc is toxic to plants at different concentrations, typical values being >200 mg/kg. The toxic levels in forage foliage are 500-1500 mg/kg. Zinc toxicity is a greater problem for plants than for animals or humans. At pH greater than 6, sludge application will not cause problems unless exceptionally high levels of Zn are added.

Organic C and N

Standard treatment methods remove most of the organic matter from raw sewage. Consequently, the typical COD, BOD, or TOC levels found in effluents are much lower than in the original sewage, and the quantities of organic C added to soils through effluent irrigation are also relatively low. For instance, the amount of organic C applied through an annual addition of 1000 mm of typical sewage effluent containing a total organic C level of 50 mg/l is only 500 kg/ha.

Since the amount of N in sewage effluents is high, the C/N ratio for the organic components is low (typical values are 5), and a release of mineral N from organic substances is favoured. The organic C/ total N ratio is much lower than the organic C/organic N ratio. Application of sewage effluent having a high BOD value reduces the oxygen level in the soil, and consequently, denitrification may be enhanced.

Soil organic matter affects soil structure and its stability, and is a source for nutrient elements essential for plant growth. The C cycle in soil consists of manifold reactions in which numerous microorganisms take place (Alexander, 1977). Many reviews deal with C transformations in the soil-plant system, some of which are specific to soils treated with organic waste (*e.g.*, Gilmour *et al.*, 1977; Stevenson, 1985).

The great concern for possible health hazards resulting from the presence of trace organics in the water is reflected in the relevant publications discussing various aspects of sewage effluent reuse. Chang and Page (1985) concluded that the risks from trace organics associated with the use of sewage effluent for irrigation are not greater than those presented by other sources of water. This is due to the reduction in the number of substances present in the effluent and in their concentration following treatment. Moreover, various soil reactions, attenuation processes, similar to those occurring in pesticide-treated soils, further reduce the level of trace organics in the soil-water system. Finally, the quantities of trace organics added through sewage effluents are usually smaller than those applied by means of standard pesticide treatment.

Biodegradation is responsible for reduction in the level of trace organics in the soil. The type of microorganisms, soil properties and the chemical properties of the organic substances involved determine the rate of decomposition. At low substrate concentrations, degradation of organic substances follows a first-order reaction. The presence of available organic C derived from other sources enhances the decomposition of trace organics in the soil, while adsorption reduces its rate.

Suspended solids in secondary effluents are mainly of biological origin. Their concentration and composition depend on the level of treatment and particularly on the degree of separation of sludge from the treated sewage in the final sedimentation process. Turbidity in excess of 20-30 mg/l (a common value for biological treatments) may be found in over/or underloaded systems. Suspended particles in secondary effluent leaving the treatment plant contain various microorganisms, such as bacteria, protozoa, rotatoria and phytoflagellata, ranging in size from colloidal to several hundred microns in size. Organic and mineralized flocs of 1 mm or even large in size (usually of porous structure), as well as clay particles and other inorganic colloidal matter, are also present.

IRRIGATION WITH WASTE WATER

Nutrients in municipal waste water and treated effluents are of particular advantage over conventional irrigation water sources and supplemental fertilizers are sometimes not necessary. However, additional environmental and health must be taken into account when treated waste water is the source of irrigation water.

Irrigation Methods

Different methods are used by farmers to irrigate crops. They range from watering individual plants from a can of water to highly automated irrigation through centrally controlled system. However, from the point of wetting the soil, these methods can be grouped under five headings, namely:

- ***Flood irrigation:*** Water is applied over the entire field to infiltrate into the soil (*e.g.* wild flooding, contour flooding, borders, basins, etc.).

- ***Furrow irrigation:*** water is applied between ridges (*e.g.* leveled and graded furrows, contour furrows, corrugations, etc.). Water reaches the ridge, where the plant roots are concentrated, by capillary action.
- ***Sprinkler irrigation:*** water is applied in the form of a spray and reaches the soil very much like rain (*e.g.* portable and solid set sprinklers, traveling sprinklers, spray guns, centre-pivot systems, etc.). The rate of application is adjusted so that it does not create ponding of water on the surface.
- ***Sub-irrigation:*** water is applied beneath the root zone in such a manner that it wets the root zone by capillary rise (*e.g.* subsurface irrigation canals, buried pipes, etc.). Deep surface canals or buried pipes are used for this purpose.
- ***Localized irrigation:*** water is applied around each plant or a group of plants so as to wet locally only the root zone (*e.g.* drip irrigation, bubblers, micro-sprinklers, etc.). The application rate is adjusted to meet evapotranspiration needs so that percolation losses are minimized.

TABLE 9.5: Basic features of some irrigation systems

Irrigation method	Topography	Crops	Remarks
Widely spaced borders	Land slopes capable of being graded to less than 1% slope and preferably 0.2%	Alfalfa and other deep rooted close-growing crops and orchards	The most desirable surface method for irrigating close-growing crops where topographical conditions are favourable. Even grade in the direction of irrigation is required on flat land and is desirable but not essential on slopes of more than 0.5%. Grade changes should be slight and reverse grades must be avoided. Cross slops is permissible when confined to differences in elevation between border strips of 6-9 cm. Water application efficiency 45-60%.

Irrigation method	Topography	Crops	Remarks
Graded contour furrows	Variable land slopes of 2-25% but preferable less	Row crops and fruit	Especially adapted to row crops on steep land, though hazardous due to possible erosion from heavy rainfall. Unsuitable for rodent-infested fields or soils that crack excessively. Actual grade in the direction of irrigation 0.5-1.5%. No grading required beyond filling gullies and removal of abrupt ridges. Water application efficiency 50-65%.
Rectangular checks (levees)	Land slopes capable of being graded so single or multiple tree basins will be leveled within 6 cm	Orchard	Especially adapted to soils that has either a relatively high or low water intake rate. May require considerable grading. Water application efficiency 40-60%.
Sub-irrigation	Smooth-flat	Shallow rooted crops such as potatoes or grass	Requires a water table, very permeable subsoil conditions and precise leveling. Very few areas adapted to this method. Water application efficiency 50-70%.
Sprinkler	Undulating 1-> 35% slope	All crops	High operation and maintenance costs. Good for rough or very sandy lands in areas of high production and good markets. Good method where power costs are low. May be the only practical method in areas of steep or rough topography. Good for high rainfall areas where only a small supplementary water supply is needed. Water application efficiency 60-70%.
Localized (drip, trickle, etc.)	Any topographic condition suitable for row crop farming	Row crops or fruit	Perforated pipe on the soil surface drips water at base of individual vegetable plants or around fruit trees. Has been successfully used in Israel with saline irrigation water. Still in development stage. Water application efficiency 75-85%.

Management Treated Waste Water

Success in using treated waste water for crop production will largely depend on adopting appropriate strategies aimed at optimizing crop yields and quality, maintaining soil productivity and safeguarding the environment. Several alternatives are available and a combination of these alternatives will offer an optimum solution for a given set of conditions. The user should have prior information on effluent supply and its quality, to ensure the formulation and adoption of an appropriate on-farm management strategy.

Basically, the components of an on-farm strategy in using treated waste water will consist of a combination of:

- Crop selection,
- Selection of irrigation method, and
- Adoption of appropriate management practices.
- Furthermore, when the farmer has additional sources of water supply, such as a limited amount of normal irrigation water, he will then have an option to use both the effluent and the conventional source of water in two ways, namely-
- By blending conventional water with treated effluent, and using the two sources in rotation.

TABLE 9.6: Evaluation of common irrigation methods in relation to the use of treated waste water

Parameters of evaluation	Furrow irrigation	Border irrigation	Sprinkler irrigation	Drip irrigation
Foliar wetting and consequent leaf damage resulting in poor yield	No foliar injury as the crop is planted on the ridge	Some bottom leaves may be affected but the damage is not so serious as to reduce yield	Severe leaf damage can occur resulting in significant yield loss	No foliar injury occurs under this method of irrigation

Parameters of evaluation	Furrow irrigation	Border irrigation	Sprinkler irrigation	Drip irrigation
Salt accumulation in the root zone with repeated applications	Salts tend to accumulate in the ridge which could harm the crop	Salts move vertically downwards and are not likely to accumulate in the root zone	Salt movement is downwards and root zone is not likely to accumulate salts	Salt movement is radial along the direction of water movement. A salt wedge is formed between drip points
Ability to maintain high soil water potential	Plants may be subject to stress between irrigations. Plants may be subject to water stress between irrigations	Not possible to maintain high soil water potential throughout the growing season	Possible to maintain high soil water potential throughout the growing season and minimize the effect of salinity	
Suitability to handle brackish waste water without significant yield loss. Fair to medium. With good management and drainage acceptable yields are possible	Fair to medium. Good irrigation and drainage practices can produce acceptable levels of yield	Poor to fair. Most crops suffer from leaf damage and yield is low	Excellent to good. Almost all crops can be grown with very little reduction in yield	

Source: Kandial (1990b)

Crop Selection

Overcoming Salinity Hazards

The ability of a crop to adjust to salinity is extremely useful. In areas where a build-up of soil salinity cannot be controlled at an acceptable concentration for the crop being grown, an alternative crop can be selected that is both more tolerant of the expected soil salinity and able to produce economic yields. There is an 8-10 fold range in the salt tolerance of agricultural crops. This wide range in tolerance allows for greater use of moderately saline water, much of which was previously thought to be unusable. It also greatly expands the acceptable range off water salinity (ECw) considered suitable for irrigation.

Selection of Irrigation Methods

Under normal conditions, the type of irrigation method selected will depend on water supply conditions, climate, soil, crops to be grown, cost of irrigation method and the ability of the farmer to manage the system. However, when using waste water as the source of irrigation other factors, such as contamination of plants and harvested product, farm workers, and the environment, and salinity and toxicity hazards, will need to be considered. There is considerable scope for reducing the undesirable effects of waste water use in irrigation through selection of appropriate irrigation methods.

The choice of irrigation method in using waste water is governed by the following factors-

- The choice of crops,
- The wetting of foliage, fruits and aerial parts,
- The distribution of water, salts and contaminants in the soil,
- The ease with which high soil water potential could be maintained,
- The efficiency of application, and
- The potential to contaminate farm workers and the environment.

A border system gives complete coverage of the soil surface with treated effluent and is normally not an efficient method of irrigation. This system will also contaminate vegetable crops growing near the ground and root crops and will expose farm workers to the effluent more than any other method. Thus, from both the health and water conservation points of view, border irrigation with waste water is not satisfactory.

Furrow irrigation, on the other hand, does not wet the entire soil surface. This method can reduce crop contamination, since plants are grown on the ridges, but complete health protection cannot be guaranteed. Contamination of farm workers is potentially medium to high, depending on automation. If the effluent is transported through pipes and delivered into individual furrows by means of gated pipes, risk to irrigation workers will be minimum.

The efficiency of surface irrigation methods in general, border, basin, and furrow, is not greatly affected by water quality, although the health risk inherent in these systems is most certainly of concern. Some problems might arise if the effluent contains large quantities of suspended solids and these settle out and restrict flow in transporting channels, gates, pipes and appurtenances. The use of primary treated sewage will overcome many of such problems. To avoid surface ponding of stagnant effluent, land leveling should be carried out carefully and appropriate land gradients should be provided.

Sprinkler, or spray, irrigation methods are generally more efficient in terms of water use since greater uniformity of application can be achieved. However, these overhead irrigation methods may contaminate ground crops, fruit trees and farm workers. In addition, pathogens contained in aerosolized effluent may be transported downwind and create a health risk to nearby residents. Generally, mechanized or automated systems have relatively high capital costs and low labour costs compared with manually-moved sprinkler systems. Rough land leveling is necessary for sprinkler systems, to prevent excessive head losses and achieve uniformity of wetting. Sprinkler systems are more affected by water quality than surface irrigation systems, primarily as a result of the clogging of orifices in sprinkler heads, potential leaf burns and phytotoxicity when water is saline and contains excessive toxic elements, and sediment accumulation in pipes, valves and distribution systems. Secondary waste water treatment has generally been found to produce an effluent suitable for distribution through sprinklers, provided that the effluent is not too saline. Further precautionary measures, such as treatment with granular filters or micro-strainers and enlargement of nozzle orifice diameters to not less than 5 mm, are often adopted.

Localized irrigation, particularly when the soil surface is covered with plastic sheeting or other mulch, uses effluent more efficiently, can often produce higher crop yields and certainly provides the greatest degree of health protection for farm workers and consumers. Trickle and drip irrigation systems are expensive, however, and require a high quality of effluent to prevent clogging of the emitters through which water is slowly released into the soil. Solids in the effluent or biological growth at the emitters will create problems but gravel filtration of secondary treated effluent and regular flushing of lines have been found to be effective in preventing such problems in Cyprus (Papadopoulos and Stylianou, 1988). Bubbler irrigation, a technique developed for the localized irrigation of tree crops avoids the need for small emitter orifices but careful setting is required for its successful application (Hillel, 1987).

When compared with other systems, the main advantages of sprinkler irrigation seem to be:

- Increased crop growth and yield achieved by optimizing the water, nutrients and air regimes in the root zone,
- High irrigation efficiency- no canopy interception, wind drift or conveyance losses and minimal drainage losses,
- Minimal contact between farm workers and effluent,
- Low energy requirements- the trickle system requires a water pressure of only 100-300 k Pa (1-3 bar),
- Low labour requirements- the trickle system can easily be automated, even to allow combined irrigation and fertilization (sometimes termed fertigation).

Apart from the high capital costs of drip irrigation system another limiting factor in their use is that they are only suited to the irrigation of row crops.

Thus, the decision on irrigation system selection will be mainly a financial one but it is to be hoped that the health risks associated with the different methods will be taken into account. The method of effluent application is one of the health control measures possible, along with crop selection, waste water treatment and human exposure control. Each measure will interact with the others and thus a decision on irrigation system selection will have an influence on waste water treatment requirements, human exposure control and crop selection (for example, row crops are dictated by trickle irrigation). At the same time the irrigation techniques feasible will depend on crop selection and the choice of irrigation system might be limited if waste water treatment has already been decided before effluent use is considered.

Management Practices

Timing of irrigation

The timing of irrigation, including irrigation frequency, pre-planting irrigation and irrigation prior to a winter or rainy season, can reduce the salinity hazard and avoid water stress between irrigations. Some of these practices are readily applicable to waste water irrigation.

In terms of meeting the water needs of crops, increasing the frequency of irrigation will be desirable as it eliminates water stress between irrigations. However, from the point of view of overall water management, this may not always produce the desired results. For example, with border, basin and other flood irrigation methods, frequent irrigations may result in an unacceptable increase in the quantity of water applied, decrease in water use efficiency and larger amounts of water to be drained. However, with sprinklers and localized irrigation methods, frequent applications with smaller amounts may not result in decrease in water use efficiency and, indeed, could help to overcome the salinity problem associated with saline irrigation water.

Pre-planting irrigation is practised in many irrigation schemes for two reasons, namely: (i) to leach salts from the soil surface which may have accumulated during the previous cropping period and to provide a salt-free environment to germinating seeds (it should be noted that for most crops, the seed germination and seedling stages are most sensitive to salinity); and (ii) to provide adequate moisture to germinating seeds and young seedlings. A common practice among growers of vegetable crops is to pre-irrigate the field before planting, since irrigation soon after planting could create local water stagnation and wet spots that are not desirable.

Treated waste water is a good source for pre-irrigation as it is normally not saline and the health hazards are practically nil.

Blending of waste water with other water

One of the options that may be available to farmers is the blending of treated sewage with conventional sources of water, canal water or ground water, if multiple sources are available. It is possible that a farmer may have saline ground water and, if he has non-saline treated waste water, could blend the two sources to obtain a blended water of acceptable salinity level. Further, by blending, the microbial quality of the resulting mixture could be superior to that of the un–blended waste water.

Alternating treated waste water with other water sources

Another strategy is to use the treated waste water alternately with the canal water or ground water, instead of blending. From the point of view of salinity control, alternate applications of the two sources will be superior to blending.

CHAPTER 10

Domestic and Industrial Wastes for Organic Farming

The use of sewage sludge and other organic materials as fertilizers for plant growth improvement is not new. The practice is almost as old as man's history on earth. Many countries have used human and animal excreta for soil fertility enhancement throughout their history. These materials have been a source of nitrogen, phosphorus, potassium, and sulphur for plants, and have contributed to the maintenance of desirable physical properties of soils, thus allowing for more abundant food and fibre production than would have been possible without them. Yet, in recent times, man has come to recognize that application of excessive quantities of these materials to land poses potential pollution problems, and a rising concern has become evident in many localities.

Concern over environmental pollution has risen greatly, causing us to evaluate more closely the effect that many of our waste disposal practices are having on our natural resources: the air, the water, and the soil. There are some contemporary aspects that tend to move sewage sludge away from a waste connotation and towards acceptance as an economic resource. World energy needs have increased due to apparent shortages of inexpensive energy sources (*e.g.*, fossil fuels). This and economic inflation have caused the cost of inorganic fertilizers to increase. As the price of inorganic fertilizers increased, the comparative value of the lower nutrient level sludge also increased, making it more competitive as a plant nutrient supplier. In addition, with a tendency towards the development of greater environmental and resource conservation conscience, society feels that it is somewhat immoral to waste valuable soil amendments by disposal of sewage sludge; thus, social pressures dictate beneficial recycling of sludge. This newly acquired social conscience with respect to the environment has urged legislatures to pass laws prohibiting disposal of sewage sludge by ocean dumping, land filling, or incineration. These legislative restrictions rapidly changed the economic considerations, since they left fewer alternatives for sludge disposal (even though there were apparent economic advantages of such disposal methods). These and other factors have forced us into considering sewage sludge for its value as a fertilizer that supplies nitrogen, phosphorus, potassium, sulphur, and several other essential plant nutrients.

It takes an average of about 10,000 people in a community to produce the equivalent of 1 tonne of dried, anaerobically digested sewage sludge per day. In a year, this would be 365 tonnes or 365,000 kg. If 3.3% of that sludge were nitrogen, that would be about 11,000 kg of nitrogen per year. If all that were available, it could fertilize from 40 to 100 ha, depending on the crop and management system.

Actually, all the nitrogen will not be available. Very likely, even with excellent management, only 40 to 50% of the total nitrogen will be available during the first year, depending on the proportion of the total nitrogen that is in the inorganic form. Therefore, possibly 20 to 40 ha of productive land could be adequately fertilized from the sludge of a community with a population of 10,000. In fact, if all the nitrogen contained in the sludge produced in the U.S.A. were available and usable, it would represent only about 4% of the total nitrogen fertilizer sold in that country.

In India, use of sewage and sludge in agricultural farms is an old practice with the first sewage farm coming up in 1895 at Ahmedabad. At present the country has around 200 sewage farms (Jurwarkar *et al.*, 1991) where the organic wastes are in use in 50,000 hectares. According to Narwal *et al.* (1990), 163 million litres of raw sewage is available per day in Haryana. In Punjab 113.4 million litres of raw sewage is generated per day. At present 17.4 million cubic metres of raw sewage is generated in urban areas of the country (Kansal, 1994).

Industries all over the world also produce large volume of effluents which should be properly disposed as they can be a source of pollution. Most of the countries lack appropriate technology for treatment and disposal of industrial wastes. Industries in Indian cities like Delhi, Kanpur and Calcutta which are located on the banks of rivers have attracted attention of the environmentalists on account of the large volume of polluting material dumped into the river water. However, effluents of some industries have agronomic value and if properly treated and applied, can improve soil productivity.

There are some negative aspects that are aptly pointed out by those who oppose sludge application to the land. These include the potential problems of pathogenic populations in sewage sludges and distribution of disease; the application of toxic levels of heavy metals that may inhibit plant growth or may enter food chains and cause lethal or sub-lethal effects to humans and/or animals; possible nitrate accumulation and leaching into ground water, eventually ending up in drinking water wells; the accumulation of polychlorinated biphenyls (PCB's) and the other toxic organic compounds that may enter food chains; and finally, excessive salt accumulation in the soil and the resultant effect on germination, emergence, and soil fertility.

TREATMENT METHODS

Methods of treatment of raw-sewage vary greatly. They consist of everything from essentially no treatment- where the raw material is used directly on the land to well digest and irradiated sludge that has been dried and bagged. Applying raw sewage to land is practiced extensively in many areas of the world. In other areas, this practice is considered unsanitary and some kind of primary and even secondary treatment is recommended or required.

Generally, the raw sludge contains a number of materials consisting of various solids and liquids. The large solids are separated by a bar screen or rack, whereas the finer materials or grit are removed by a grit removal system. From here the flow stream takes the sewage to a primary settling tank or clarifier. In this tank, the solids with a density greater than that of the water settle on the bottom, while the solids that float are skimmed off the top. Both of these groups of solid materials are then sent to either an aerobic digester or an anaerobic digester.

The water that is separated can be piped to a trickle filter or to an activated sludge tank. As it passes through the trickle filter or activated sludge tank, many of the soluble and/or biodegradable organic compounds are converted into biomass and CO_2. Following either of these treatments, the solid biomass is separated from the water in another clarifier, whereupon the water is chlorinated and released to a stream.

The solids or sludge from either the top or the bottom of the primary settling clarifier can be piped to anaerobic or aerobic digesters for further processing. The solids from the clarifier separation following the trickle filter may be sent either to an activated sludge tank or directly to the digesters. Solids from the activated sludge tank clarifier are sent directly to the anaerobic or aerobic digesters.

The products produced with secondary digestion will be as variable as the content of the raw sewage supplied to the treatment plant. The solid contents typically vary from less than 2% up to about 6%. The relatively stable digested sludge can be handled in several possible ways when land application and recycling are considered as a means of utilization. Without further treatment, the sludge can be applied to a field soil- on the surface, using flood or furrow irrigation, or by sprinkler or dribbled distribution systems. It also can be incorporated underground with subsoil injection equipment. If injected, several potential problems, such as nitrogen loss, pathogen exposure, and odour generation, are partially if not completely eliminated (which is obviously impossible in surface application).

One of the biggest problems associated with using the products of the secondary digester directly for land application is the high cost of transporting large quantities of water by tank truck or pipe pumping to use small quantities of solids on the land. Therefore, some treatment plants de-water or dry their sludges. This can be done in several ways, including sand bed filtration, vacuum or pressure filtration, and centrifugation. Sand bed filtration and drying is a slow process. It may take six weeks to two or three months, but the product is relatively dry, ranging from 40 to 60%

solids, is easily handled with a pitch fork or shovel, and can be stored in piles or hauled to the field by open bed truck and spread on the land. Some communities are heating or irradiating and bagging this material for use as fertilizers and soil conditioners.

Filter cake, a material that is about 15 to 20% solids and has the appearance of a thick mud, is a result of vacuum filtration, pressure filtration, or centrifugation. Flocculation of the solids is required for successful vacuum or pressure filtration. This can be accomplished by adding iron chloride and lime or by adding a number of suitable polymers presently available. The filter cake may be spread on the land surface or it can be injected underground using screw-type pumps which can handle mud-like materials. Filter cake is also disposed of via landfill. The composted product is separated from the wood chips and used as a fertilizer and soil conditioner. The filter cake can be spread on the land surface and incorporated into the soil with tillage equipment or it can be injected underground using the screw type pumps.

CHARACTERISTICS OF WASTES

Like most products of biological systems, sewage sludge varies greatly in chemical, biological, and physical properties. These characteristics will depend on the sources of sewage, the treatment system, and the management of the treatment system. Sludges from industrialized areas with laboratories and chemical processing facilities will generally have greater heavy metal contents than sludges from residential areas, even though zinc and copper originating in the plumbing pipes of homes are added to sewage systems of all modern communities. It is of interest to note that the sewage from homes varies in chemical composition with time of day. The waters in sewers in the morning are higher in zinc and copper than at other times of the day, resulting from water being stationary in the plumbing pipes at night, causing zinc and copper to dissolve slightly. Sludge composition changes with season of the year, especially when there are seasonal industries in the community. For example, pickle processing plants use large quantities of sodium chloride or brine. As the used brine is discarded into the sewer system, the content of sodium and chloride is greatly increased. Other industries use other ingredients that may be undesirable in sewage sludge when land application is being considered. The differences between the minimum and maximum concentrations of most elements are from less than one to many orders of magnitude.

TABLE 10.1: Elemental composition of sewage sludge

Component	Minimum	Maximum	Median
	Concentration (%)		
Organic C	6.5	48.0	30.4
Inorganic C	0.3	43.0	1.4
Total N	<0.1	17.6	3.3
NH_4^+-N	<0.1	6.7	1.0
NO_3^--N	<0.1	0.5	<0.1
Total P	<0.1	14.3	2.3
Inorganic P	<0.1	2.4	1.6
Total S	0.6	1.5	1.1
Ca	0.10	25.0	3.9
Fe	<0.10	15.3	1.1
Al	0.10	13.5	0.4
Na	0.01	3.1	0.2
K	0.02	2.6	0.3
Mg	0.03	2.0	0.4
	Concentration (ppm)		
Zn	101	27,800	1,740
Cu	84	10,400	850
Ni	2	3,515	82
Cr	10	99,000	890
Mn	18	7,100	260
Cd	3	3,410	16
Pb	13	19,730	500
Hg	<1	10,600	5
Co	1	18	4
Mo	5	39	30
Ba	21	8,980	162
As	6	230	10
B	4	757	33

Source: Dowdy et al. (1976)

The composition of domestic effluent mainly depends upon the source, method of collection and treatment provided. In many countries, including India sewage is a mixture of both domestic and industrial wastes. Table 10.1 shows the composition of sewage sludge on the basis of 200 samples from different states of USA (Dowdy *et al.,* 1976). Several scientists have reported that the sewage water of Indian cities and towns is a rich source of essential plant nutrients (Goel and Trivedi, 1984; Singh and Kansal, 1985a; Kansal, 1992). Singh and Kansal (1985a) analysed samples from various towns of Punjab (India) and reported appreciable quantities of N, P, K, S and Ca.
The nutrient content varied from town to town depending upon the sources *i.e.* industrial wastes or atmospheric fall out. Tiwana *et al.* (1987) analysed waste water samples from Ludhiana city and observed very high concentration of Nickel, hexavalent chromium and cyanides. The content of cyanide and chromium of these samples were one thousand times higher than that of New York.

In many towns and cities, the farmers use the sewage sludge directly in their fields as a source of irrigation or supplementing the essential nutrients. The disposal of sewage sludge in water bodies deteriorates their quality in respect of pH, EC. And especially the trace elements which at times lead to toxicity. Juwarkar *et al.* (1991) did not observe high concentration of toxic elements except for Cd for Jaipur, Madras and Kanpur and Ni in the Kanpur and Delhi (Okhla) sewage sludge (Table 10.2).

Not only do sludges vary greatly in chemical composition, but they vary in physical properties, too, depending on the method of treatment and management. Some sludge has as little as 1 or 2% solids carried in a water suspension. Typical of anaerobically digested sludge is a solid content of 4 to 5%. This material can be applied satisfactorily to soils by open furrow methods, sprinklers, and dribble systems, or by underground injection. Transportation costs become prohibitive if this type of sludge has to be transported great distances from the treatment facility, because of the large quantities of water involved. In order to overcome this problem, many treatment plants vacuum filter their sludge to de-water from 95 to 96% water down to 80 to 85% or less. This product is a mud-like material that can be transported in an open box truck rather than a tank truck.

TABLE 10.2: Characteristics of sewage sludge from Indian cities

City		Heavy metals (mg/g)			
		Cd	Cr	Ni	Pb
Allahabad	T	3.5	60.4	32.3	76.8
	A	0.8	-	2.3	8.3
Delhi (Okhla)	T	1.5	82.0	191.5	41.7
	A	0.5	-	7.6	6.7
Madras	T	8.3	38.5	60.5	16.5
	A	1.3	-	2.2	1.9
Jaipur	T	7.3	176.2	37.5	66.0
	A	1.4	-	0.3	6.4

T = Total, A = Available:
Source: Juwarkar et al. (1991)

The de-watered material will have high concentrations of iron chloride and lime, a polymer, or a mixture of both since these are needed to flocculate the sludge before vacuum filtering. The addition of these chemicals and the filtration process not only change the physical condition but also change the biological and chemical nature of the resulting so called filter cake. Whereas the unfiltered sludge has a considerable population of organisms (pathogenic and non-pathogenic), the lime and iron chloride treated filter cake has a very low population due to the extremely high pH of the sludge.

WASTE DECOMPOSITION IN SOILS

Unlike fresh plant and animal residues that are incorporated into the soil, most sewage sludges have been through a biological treatment, wherein partial decomposition and stabilization have occurred. This results in a decrease in total amount of organic material, due to loss of carbon, either as carbon dioxide in an aerobic environment, or as methane and carbon dioxide in an anaerobic digestion. When the sludge is added to soil, the rate of decomposition will be somewhat slower than that of most fresh organic residues since the available energy material (soluble organic carbon compounds) and nutrients for microbial growth are less. The result is an increase in the level of soil organic matter, sustained over a longer period of time. In a given environment, this may have its advantages, including an increased source of plant nutrients and improved air and water relationships, resulting in greater plant growth. However, the possible increased mobility of some of the heavy metals due to the addition of naturally occurring chelating compounds in the sludge may cause leaching of the chelated metals into the ground water.

This specific decomposition rate of many plant and animal residues in soils have been determined, but little data are available on the rates of decomposition of sewage sludges added to soils. Estimates have ranged from 2% to greater than 60%. Rothwell and Hortenstine (1969) and Miller (1973a) determined that 17 to 20% of the organic carbon was mineralized in a six month laboratory incubation study, but Larson *et al* (1972) determined that 20 to 28% of the sewage sludge added to a soil in a laboratory incubation study decomposed in a 365 day period at typical room temperature (21°C ± 2), based on CO_2 production. Generally, the higher the proportion of the sludge in the soil (higher application rate), the lower the decomposition percentage was in a given time period.

Sewage sludge must be considered a low analysis fertilizer and soil conditioner. Sewage sludges range from less than 0.1 to 17.6% nitrogen (with a median value of 3.3), less than 0.1 to 14.3% phosphorus (with a median value of 2.3) and less than 0.02 to 2.6% of potassium (with a median value of 0.3). Other nutrient elements are contained in greater or lesser amounts but are generally of less concern fertility-wise than the above mentioned elements. Only a portion of each of the essential nutrients contained in sewage sludge is available for plant growth during the first plant growing season after application. Those elements that are an integral part of the organic molecule in sludge are not immediately available to plant growth until mineralization (decomposition of the organic molecules to release inorganic ions) occurs. However, most of the nutrients that are found in organic molecules- such as nitrogen, phosphorus, and sulphur are usable by the plants. For example, the median total nitrogen figure is 3.3%, whereas the median inorganic nitrogen value is about 1.0%. Thus, potentially 30% of the total nitrogen is immediately available for plant growth. According to the data for phosphorus about 70% is in the organic form, some of which will be available for plant use during the first growing season after sludge application. Most of the potassium will be in the inorganic form, and will be available to the plants immediately. Although some other elements contained in sewage sludge are essential for plant growth, adding sludge as a fertilizer to supply adequate amounts of most of them is not considered important except in areas where those specific nutrient deficiencies are encountered.

Of greater concern in the management of sludge application as a fertilizer to supply nitrogen, phosphorus, and potassium, is the concomitant addition of excessive amounts of elements such as zinc, copper, lead, nickel, and cadmium that can accumulate in the soil to toxic levels- toxic to either the plants that grow in the soil or to the animals (including man) that eat the plants. The possibility of toxic levels of nitrate, heavy metals, or toxic organic compounds entering into food chains represents one of the greatest deterrents to acceptance of land application of sewage sludge as an effective recycling practice.

Nitrogen

Most studies have not been able to distinguish between the NH4+ and NO3– that originated from the soil organic matter and the sludge. One method of approach to this difficulty would be to feed humans foods isotopically labeled with N15, then digest the faecal wastes in an anaerobic digester, after which the resulting sludge could be added to soil and the rate of mineralization of the N15 labeled organic nitrogen compounds measured. This would allow one to distinguish between the inorganic nitrogen coming from the soil organic matter and that coming from the sludge. Other alternatives would include adding highly enriched N15 to an anaerobic digester to which raw sewage sludge is gradually added, thus allowing the production of N15 enriched sludge. This has been about 2% N15, which is further diluted as the sludge is mixed with the soil, thus making precise measurements somewhat difficult.

An interesting phenomenon relating to this problem is the so-called priming effect that the addition of sewage sludge has on the native soil organic matter decomposition rates. A decomposition rate of 0.5 to 2% of the native soil organic matter may be increased considerably by addition of sewage sludge, thus releasing extra nitrogen for plant growth beyond what is normally anticipated from the soil organic matter. This further complicates the problem of recommending an accurate and sound application rate for sewage sludge.

TABLE 10.3: Plant growth and nitrogen uptake in relation to added sludge and fertilizer-N

Application Rate	Plant Growth (g/pot)	N Uptake by Plants (mg/pot)
Sewage Sludge (t/ha)		
– 0	1.8	45
– 22.4	4.6	212
– 56	8.9	341
– 112	8.3	335
– 224	6.0	265
Fertilizer-N (kg/ha)		
– 0	1.8	45
– 67	3.0	65
– 134	3.7	100
– 269	5.3	170

Source: Sabey (1977)

Several aspects of the nitrogen cycle are of significance to environmentally sound and successful management of a land sludge recycling system. If we do not want to pollute the ground water by sludge or effluent applications to soil, we must either add only enough to supply the nitrogen for plant growth or we must learn to manipulate the cycle to decrease nitrogen in the soil-sludge mixture. This latter alternative is possible at two places in the cycle. We can enhance ammonia volatilization or enhance denitrification. Ammonia volatilization can be promoted by raising the pH, spraying, or drying the material before incorporating it into the soil. Denitrification can be increased in soils by adding readily available sources of energy material, such as sugars, alcohols, or other easily oxidizable carbon compounds, and by depleting the oxygen by saturating the soil with water after nitrite or nitrate has been produced.

When evaluating use of sewage sludge as a supplier of nutrients, the addition of excessive sewage sludge is not often considered, so mechanisms to promote denitrification or volatilization are ignored, although these mechanisms will occur naturally in the presence of excess sludge, so an understanding of them is pertinent.

It appears that the equivalent nitrogen fertilizer values for 0, 4, 8, 20 and over 40 tonnes of sludge per hectare are 0, 167, 458, 410 and 263 kg/ha, respectively, or 287, 502, 490 and 375 inorganic nitrogen is made on the basis of plant growth or on the basis of nitrogen uptake. These values may be quite valid for the soil, the plant used, and the other conditions of this particular study, but care must be exercised in extrapolating to other soils, crops, and conditions.

Phosphorus

Even though many sludges contain less than half as much phosphorus as nitrogen, there are not as many ways to promote phosphorus losses from the soil as there are for nitrogen. Short of incineration, little or no phosphorus volatilization occurs in soils, except possibly phosphine (PH3) in paddy soils. However, because of the large amounts of aluminium, iron, and calcium they contain, many medium to fine textured soils have an almost unlimited capacity to convert phosphorus into nearly insoluble compounds with these elements. It is therefore doubtful that on these soils, phosphorus will limit sewage sludge applications.

Potassium

The quantities of potassium in most sewage sludges are quite low, therefore, when using sludge as a fertilizer for soils that are deficient in potassium, it may be necessary to add supplemental potassium. However, in arid areas, many of which have soils with ample available potassium, some sewage sludges make a balanced fertilizer, supplying ample nitrogen and phosphorus as well as some of the micro-nutrients.

Essentially, all of the potassium contained in sewage sludge will be in the inorganic form; therefore, when added to the soil, it does not have to undergo mineralization to become available to plants, as is true of much of the nitrogen and phosphorus. Some potassium fixation can occur, however, especially in the presence of soils containing clays such as smectites and hydrous micas with a 2:1 crystal structure.

Micro-nutrients and Heavy Metals

In the past, concern for micro-nutrients, including the essential heavy metals such as copper and zinc, has been in the area of deficiencies, and questions have not arisen concerning excessive amounts that could be toxic to either plants or the animals that ingest the plants. With respect to sludge application, concern has turned to toxicities and how much of thc hcavy mctals, including cadmium, nickel, lead, and chromium, as well as zinc and copper, can be added before the plants are detrimentally affected and toxic levels accumulate in food chains. Most municipal sludges contain appreciable quantities of copper, zinc, iron, and manganese, with lesser quantities of boron, molybdenum, and chlorides. Generally, when sludges are added to most soils at rates supplying adequate nitrogen, they supply ample quantities of these micro-nutrients for plant growth. Additionally, the naturally occurring chelating compounds contained in-or produced during the decomposition of the sludge tend to render these micronutrients more available to plants. The solubility and mobility of the metals added to the soil with sludge also vary with soil conditions and organism activity.

Soil micro-organisms not only affect the oxidation-reduction potential that influences solubility of the metals but also produce organic chelates that can result in leaching of the metals into the ground water, especially in acid soils. Metals are generally more soluble and mobile in low pH than in high pH soils and this fact makes it possible to either decrease or increase solubility, plant availability, and leachability of metal ions as the situation warrants.

ASSOCIATED PROBLEMS

Nitrate Accumulation

The accumulation of nitrate in ground water, well water, lakes, and streams is considered by many to be undesirable. 10 parts per million of nitrate nitrogen is the maximum permissible concentration in water used for human consumption. One of the reasons for such a guideline is the fear of methemoglobinemia (Blue Baby Disease) when water containing a high concentration of nitrate is used for baby formulas. Additionally, high nitrate levels have been cited as one of the causes of algal bloom in streams and eutrophication of lakes. Ryther and Dunstan (1971), have implicated fertilization of agricultural lands as a major cause of higher nitrate levels in surrounding lakes and streams. Others have indicated that there is enough nitrogen in rainwater and in most lakes to fully support algae growth. Unless the nitrate level is too low for algae growth, the addition of more nitrate will not trigger algal bloom.

A mechanism that has considerable potential for triggering algal bloom is the addition of organic debris to water bodies. From the decomposing organic matter, carbon dioxide is produced. This carbon dioxide, along with water and sunlight, allows for rapid algae growth if sufficient levels of other essential elements are present. Most elements, including nitrogen and phosphorus, are contained in most water bodies and streams in high enough concentrations (0.01 parts per million phosphorus, 0.3 parts per million nitrogen) for algal growth. The extent to which sewage sludge applications will contribute to nitrate pollution will largely be dependent on management. If only sufficient sludge is added to supply adequate nitrogen for plant growth, there is unlikely to be a problem of nitrate pollution. Even when amounts of sludge are excessive of that needed for plant nutrition, nitrate pollution may not occur if denitrification can be promoted. This would convert much of the leachable nitrate to nitrogen oxide gases or to nitrogen gas, both of which would diffuse back into the atmosphere. Many sewage treatment plant operations are interested in disposing of the sewage sludge produced by their plants. There is, therefore, a tendency to apply as much sludge per unit area of land as possible. With this concept in mind, excessive amounts of sludge will be added to the field. This can be illustrated by the following example. Assume that 47 tonnes/ha of dry, anaerobically digested sludge is added to a field. The total nitrogen contained in the sludge could be about 1 tonne; of the 1,000 kg of nitrogen about 300 to 400 kg could be ammonium and nitrate nitrogen. This inorganic nitrogen is immediately available for plant utilization unless it is lost by ammonia volatilization during drying at the soil surface or by denitrification once nitrate is formed. Most plants will use no more than about 225 to 340 kg/ha of nitrogen. If no losses occurred, this would leave 75 to 175 kg of ammonium nitrogen that would eventually be converted to nitrate and have a high potential of being leached into ground water, wells, lakes, or streams. Fortunately, there will most likely be some losses occurring during or soon after surface application and during the decomposition process, especially if the system is managed with this objective in mind. In addition to the available inorganic nitrogen, some of the organic nitrogen will be mineralized and supply potentially leachable nitrate. At 67 tonnes/ha, the inorganic nitrogen contained in the sludge or produced during decomposition is excessive compared to that needed for optimum plant growth. Unless management practices are designed for removal of the excess inorganic nitrogen, ground and well-water pollution or runoff pollution are feasible.

Phosphorus Accumulation

Phosphorus differs markedly from nitrogen in several ways that relate to its pollution potential. First, phosphorus in its various forms is much less mobile than nitrate. The tendency to leach through the soil is much less. Second, phosphorus is not as easily lost from the soil by volatilization as is nitrogen. Third, phosphorus is used by plants to a lesser extent than nitrogen. All these facts lead to the conclusion that phosphorus could accumulate in the soil if sewage sludge were added in excessive quantities. The major removal mechanisms are by plant growth and erosion caused by surface runoff. When sludge application rates are governed by nitrogen requirements and when the phosphorus content of sludge is typically about one-half that of nitrogen, excessive phosphorus will be added to the soil. Fortunately, much of the phosphorus is converted to insoluble forms in many soils so it may not present a problem in those soils. In other soils, however, an overabundance of phosphorus can cause zinc deficiencies in plants, as well as other problems. After some period of time of sludge application, phosphorus may have to be used as the element controlling sludge application rate instead of nitrogen.

Heavy Metal Accumulation

Many sewage sludges, especially from highly industrialized communities, are laden with high concentrations of heavy metals, each of which can be potentially toxic to plants, animals, or humans. The threat of sub-lethal effects of heavy metals is of concern to many people. The possibility of lethal effects to plants growing on soils which have been treated with sludge for many years is more likely, but even this has not been fully documented, so irrefutable proof of potential serious problems of heavy metal accumulation and uptake on sludge applied land has yet to be established. Plants vary greatly in their tendency to concentrate metals. The green leafy vegetables such as, beet and lettuce are excellent concentrators of the metals, whereas most grains are less so. Additionally, various plant parts differ greatly in their concentration of the metals, with the leafy portions concentrating more than the stem and seed portions. An interesting aspect of the problem is that some metals which can be accumulated by plants can be toxic to the plant before the concentrations in the plant become toxic to humans eating the plant. Such is the case with zinc and copper, but it has been shown that cadmium can be concentrated in the plant sufficiently to become toxic to humans without being lethal to the plant. This makes cadmium a metal of great concern in considering sludge application to land for food production; thus, application rate based on cadmium content is becoming a primary or at least secondary guideline along with nitrogen and phosphorus contents.

Other metals, though of some concern, are considered much less of a problem than cadmium. Chromium is not readily taken up by plants because it is generally found in sludge in the Cr^{+++} form, which is not generally available to plants. Lead can be toxic in soils with very low phosphate levels, but in the presence of phosphate, lead is generally rendered unavailable to plants. Mercury and other metals are not typically found in sludges at high concentrations or are not soluble enough to be a serious problem when the sludges are added to the soil. This is especially true in soils containing high levels of organic matter. The organic matter increases the cation exchange of the soil, thereby increasing the tendency for metals to be adsorbed. The increase in organic matter in sludge also adds chelating compounds, some of which form stable complexes with the metals and withhold the metals from the plants. The extent of this mechanism depends on the stability constants of the chelate-metal complexes, the concentration of the chelate and the metals, and the pH of the amended soil mixture.

Pesticide Residues

Little work has been done on the effects of some of the toxic organic compounds that are contained in sewage sludges. These include insecticides, herbicides, and other pesticides, along with polychlorinated biphenyls (PCB's). It is assumed that although sewage sludges contain low concentrations of many of these pesticides, the effects would not approach those of direct spraying on the crops grown on the land. However, the long-term effects need to be investigated and considered in designing land spreading systems.

Several workers have investigated the chemical and biological stability of PCB's in soils and found them generally to be quite persistent and toxic. Tucker *et al.* (1975) reported that the persistence and solubility of the PCB's are related to the degree of chlorination.

When sludge containing PCB's is added to soils, the PCB's are absorbed both by the clay minerals and the organic matter and generally have low solubilities. However, the potential for plant uptake and leaching into the ground water is ever-present. The information on absorption of the PCB's by the plant roots or the leaves after volatilization from the soil, the effective adsorption by the soil, and the amount of leaching in the ground water, is meagre. The amount of PCB uptake by plants cannot be estimated accurately from the amount of the PCB's in the soil because many variable factors influence plant uptake. Undoubtedly, these mechanisms are dependent on many environmental factors that need further research to elucidate the many ramifications associated with this problem.

Salt Accumulation

Some sewage sludges contain considerable quantities of salts. If such sludges are applied to poorly drained lands where the water table is near the soil surface and the net water movement is upward, a salt problem could arise, particularly if concentrations of sodium salts (especially carbonate) are present. The difficulty associated with salt accumulation in the soil revolves around increased osmotic pressure in the soil solution, which may decrease plant uptake or seed germination and produce a degradation of the soil physical properties. A basic understanding of the mechanisms surrounding this problem exists, and if knowledge of the initial salt content of the soil and the sludge, the moisture movement relationships of the soil, and the quantities of water to be added to the soil is available, the magnitude of the salt problem can be predicted. In well drained irrigated soils, soil salinity is not likely to be a problem, although salts leached out of the zone of incorporation of the sludge may move through the soil profile and into the water table and degrade ground water quality.

Pathogens

Pathogenic organisms added to soil when sewage sludges are applied to land present a potential hazard to human and animal health. Most sludge is not pathogen-free, even though the numbers of most species are much lower in digested sludges than in raw or primary sludges. Some methods of treating sewage sludges prior to application to land help to decrease the hazards. These include composting, limiting, irradiation, chlorination, pasteurization, and heat-drying. Although these treatments decrease the population of pathogens, it is questionable that even these, on a practical ongoing basis, will render sludges 100% free of disease organisms. Fortunately, most of the pathogens die-off quite rapidly upon being incorporated into the soil. However, some species persist for rather long time periods by various protective mechanisms. Many factors affect the persistence of pathogens in soils, including available energy supply, moisture, temperature, aeration, soil pH, and competition with the native population. The degree of seriousness and hazard associated with applying well digested sewage sludge is not fully known.

The fact is that most, if not all, serious disease outbreaks that have been traced to sewage wastes have been due to raw or poorly digested sewage rather than too digested, well treated sludge. Burge and Marsh (1978) stated that salmonellosis in the United States has infrequently been associated with sewage sludge even though salmonella organisms are found in digested sewage sludge. They further state cases of enteric viral disease arising from well treated sewage are relatively few considering the estimates of these organisms in sludge.

Well digested sewage sludges have far fewer pathogens than do raw sewage sludges, although they are not pathogen-free. Some pathogens can survive the digestion process in the sewage treatment plant. Generally, during the time after sludge addition to soil, the pathogenic population decreases quite rapidly. Disappearance of these organisms is caused by antagonistic effects of saprophytic soil organisms, predation, nutritional factors, or other adverse environmental conditions. If the soil has a healthy, active native population, the rate of pathogen die-off is rapid.

Klein and Casida (1967) reported a reduction of 90% of *E. coli* numbers within five days in unamended soils, but if glucose was added, the lives of these organisms were prolonged in the soils. If one assumes that salmonella and related organisms act similarly, there is a possibility that pathogenic organisms found in sewage effluent and sludge and in animal manures may escape from soil in drainage water before being inactivated by soil. Peterson (1970) showed that there was a rapid decrease in fecal coliforms with lateral movement in ground water even in sandy soils. However, in test wells on the sludge application sites, fecal coliforms were detected, especially after a heavy rain. Organism movement into the well was due to gravitational water movement. In a test well, off the sludge application site, no test organisms were detected, indicating the filtering effect of soils as ground water moved laterally.

Although there is a rapid decrease with time in pathogens added to soil via effluents and sludge, some organisms have the capacity to survive in the soil by spore formation or other protective mechanisms. However, though the potential for disease is there, many persons work around sewage sludge and effluents daily with few documented cases of illness due to the pathogens contained in the wastes. With reasonably low rates and infrequent applications of wastes, the possibility of problems caused by pathogens will be small.

Viruses present a more difficult problem to handle, since culturing and isolating them is more complicated than with many other organisms. Most work on pathogens in the soil due to sewage effluents or manure application has been with bacteria. Merrill and Ward (1968) reported an absence of viruses after 1500 feet of soil percolation, but presence of viruses when the distance was only 400 feet.

MANAGEMENT OF WASTES

Application Methods

A number of methods of application of sewage sludge to land exist, each with advantages and disadvantages. The moisture condition of sludge to be spread controls, to some degree, the techniques used. Sludges that have less than 70% water are relatively dry and can be spread with a manure spreader or equipment similar to it. Sludges with moisture contents of 75 to 85% (so-called filter cake) can be hauled in an open box truck and be dumped or spread with some success with a manure spreader, especially at the drier end of the range. There are screw-type pumps that can push this material through pipes behind sweeps or in furrows into the soil. High moisture sludges ranging from 90 to 99% moisture can be spread by pump and sprinkler or dribble systems, including adapted pivot sprinkler or other self-propelled systems. The traveling sprinkler gun has been used successfully for this kind of sludge. Special provisions are usually necessary to ensure that larger chunks of solids do not clog the outlets in these distribution systems. These require pump power and pipe conveyance. Open ditch conveyance is possible with high moisture sludges. The ditch feeds a ridge and furrow or flood irrigation system in order to evenly distribute the sludge over the land. This type of sludge can also be distributed by tank truck spreading on the soil. One of the problems associated with high moisture sludges is the tendency for solids to settle to the bottom of holding lagoons and transport and feed tanks, thereby resulting in heterogeneous distribution of the solids on the land.

Application Rates

One of the problems associated with land spreading of sewage sludge is the lack of understanding of the physical, biological, and chemical reactions in the soil and thus the inability to make accurate recommendations for application rates. Several criteria have been suggested, including nitrogen and phosphorus supply for plant growth and heavy metal loadings. Nitrogen supply has been the most persistent criterion suggested by most experts. This is because of the possibility of nitrate accumulation over and above the nitrogen needs of the plant and thus the potential for nitrate leaching into the ground waters. This it is felt, should be avoided if at all possible. However, since the amount of nitrate that accumulates in the soil is dependent on several biochemical reactions, each of which is dependent on a variety of environmental conditions, nitrate accumulation and leaching potential is extremely difficult to predict.

Nutrient Availability

It is important to know the amount of total organic and inorganic nitrogen in the sludge as well as the nitrogen supplying capacity of the soil. The latter can be estimated by soil testing procedures and the former from standard chemical analysis. There are other factors that are not so easily determined or estimated. These include an estimate of the amount of ammonium that will be lost by volatilization when added in a high liquid sludge on the surface of the soil or plant tissue. This is highly dependent on the pH at which the sludge dries. At a high pH, more ammonia will be lost by volatilization than at low pH. It further requires an estimate of the amount of denitrification that occurs. Anaerobic conditions in the presence of ample available organic material are conducive to rapid loss of nitrogen into the atmosphere, resulting in less nitrate remaining in the soil. It also requires an estimate of the amount of organic nitrogen that will be mineralized in a growing season.

This depends on the stability of the sludge and the environmental conditions that influence microbial activity in the soil. If all these estimates were available, it would be relatively simple to make accurate recommendations for the proper sludge application rates necessary to supply the nitrogen needed for plant growth and avoid applications of excess nitrogen that could be leached into the ground water.

Different methods of application result in different percentages of ammonia nitrogen in the sludge that will be available to the crop being grown.

- Surface application of high liquid sludge without incorporation results in an estimated 80% loss of ammonia nitrogen, leaving about 20% of ammonia available for plant growth.
- Surface application of high liquid sludge with immediate incorporation results in 30 to 70% loss of ammonia nitrogen, depending on how soon incorporation occurs.
- Subsurface application of high liquid sludge results in negligible loss, so most of the ammonia nitrogen would be available.
- Some losses of nitrate can occur when denitrification proceeds as soils are aerated and then flooded for a period of several days. The amount depends on the conditions in the soil.

Phosphorus

The phosphorus added to the soil through sewage sludge is not generally lost from the soil as is nitrogen, although much of it can become unavailable due to precipitation or adsorption. Typically, when sludge is added to the soil to supply sufficient nitrogen, adequate phosphorus will be added for optimum plant growth, since plants need only about 1/10 to 1/5 as much phosphorus as nitrogen, and sludges typically contain about 1/2 as much phosphorus as nitrogen. The primary phosphorus removal mechanism form the soil is plant utilization and harvest from the land. Therefore, phosphorus could potentially accumulate in some soils to toxic levels and may eventually become the governing factor for sludge loading rates. After applying sludge to the land for five years or more, the phosphorus content of the soil should be monitored. If the phosphorus content exceeds 450 kg/ha, then the sludge application rates should be reduced or discontinued, as the case may be.

Potassium

The potassium content of most sludge is quite low compared to the levels of nitrogen and phosphorus. Since potassium is needed by plants in rather large amounts, a supplemental source of potassium may have to be supplied to provide a balanced nutrient level for optimum plant growth. This will be especially true in soils of low pH or soils of the more humid and highly leached parts of the world. Soils of and regions are generally rich in available K and supplemental potassium is not needed.

Micronutrients

The disposal of the effluent on agricultural land also contributes substantially in adding the available micronutrients and trace elements in the soils. Long term application of sewage water considerably increased the available content of Zn, Cu, Fe, Mn and Ni in soils around Jalandhar and Ludhiana in Punjab (Table 10.4).

TABLE 10.4: Effect of sewage irrigation on accumulation of available elements in surface soils

Location	Irrigation Source	DTPA extractable elements				
		Fe	Mn	Zn	Cu	Ni
Ludhiana	Tubewell	31.7	20.2	4.5	5.8	0.19
	Sewage	51.6	24.6	20.6	23.9	1.14
Jalandhar	Tubewell	26.4	15.2	4.9	4.3	0.20
	Sewage	40.6	25.2	21.3	25.2	0.98

Source: Kumar (1992)

From a short-term perspective, the loading rates calculated to provide nitrogen for growing crops usually result in the application of trace elements to the soil in amounts that would not result in toxic levels in water and food chains. In the long run. The trace element content of the soil should be monitored to ensure that these trace elements do not build up in the soil. The frequency of monitoring of the soil will be dependent upon the quality of the sludge. Monitoring of the trace elements in the sludge soil mixture should be done at least once every five years starting with the fifth year of continuous application.

Most of the trace elements in sludge applied to soils are retained in surface soils; only small amounts may go into solution. Thus, pollution potential for surface waters is greater than that of ground water. Since trace elements applied to the soil are largely concentrated in the surface, runoff into surface waters may contribute to the pollution of water bodies. concentration in water of trace elements considered to be toxic to aquatic organisms are, in many cases, lower than those considered toxic to higher plants and animals. Since these tolerances are so low, surface runoff of either sediment or solution into surface waters must be held to a minimum.

Optimum Conditions for Use of Wastes

It is difficult, to specify exactly the soil properties that are optimum for land application of sludge as a fertilizer. There are so many factors that influence so many aspects of the practice and some are incongruous with others. As a generalization, however, we could say that for optimum soil fertility, the extremes should be avoided. If a soil is too sandy, the infiltration may be too rapid and the adsorptive capacity may be too low. Therefore, positively charged ions and particles. As well as pathogens, may move with percolating waters to a greater extent than is desirable. On the other hand, if a soil is too high in clay, it may be poorly drained and runoff and erosion could occur.

CHAPTER 11

Future Trends in Organic Farming

The major trends in structural change in agriculture during the past four decades are well known. These structural changes have been mainly directed towards meeting the demands of the growing population, especially in the under developed and developing countries. In this process, agriculture became less organic in its technological and practical aspects as high input use played a key role in augmenting food production. Presently, agriculture is at the cross roads with higher production envisaged to meet the growing demand for food commodities and emphasis on conserving the natural resources in order to keep the earth green and healthy. Man should gear himself to address the following challenges in developing technologies relevant to organic agriculture and attaining desired production targets in future.

AGRICULTURAL TECHNOLOGY AND PRODUCTIVITY: FUTURE CHALLENGES

Natural Resources

Policy actions create rights and obligations concerning natural resources as production inputs in agriculture, residential use, commercial and industrial use, public parks, private recreation, water-sheds, and state/federal range and forest uses. Who should/can have access to these resources, for what (beneficial) use, at whose expense, and for how long are the basis of policy debates. In some cases, policy actions are taken to deliberately affect resource use in the food system- e.g., farmland zoning. Other actions, such as depreciation allowances for irrigation equipment, inadvertently affect the resource/food system relationship.

Land issues involve decisions about the amount and location of land available for food production. Almost every country has them on its policy agenda. Some countries have had farmland protection policies; townships, counties, and other local units of government have also taken action.

Soil erosion is the basis for many resource issues. This is because soil is displaced, thereby changing the structure, fertility, permeability, organic content, and other features of the land left for production. The displaced soil may simply go to increase the productivity of a neighbour's farm or, more often, to cause problems by clogging rivers, streams, or harbours with sediment and debris. National land conservation policies address the degree to which government should influence a farmer's efforts to reduce soil loss. Although landowners have some economic incentive to protect their land, society has decided that for various reasons farmers conserve too little and therefore need greater incentives. Perhaps farmers do not cope fully with erosion because its impacts can be offset for them by fertilizers or new plant varieties. Inflation or land location and development potential may also mask the impacts of erosion on land value and underscore the lack of an economic incentive enough for individual farmers to do anything about it. Soil conservation policies have been enacted to generate more conservation than farmers are willing to buy on their own.

Water quantity issues focus on who can actually use water as it is distributed geographically through time. Water quality is also an integral and critical issue, since it is inextricably bound to water quantity considerations.

Soil Conservation

As farmers ploughed under more land to take advantage of rising prices, old conservation practices were lost. Narrow terraces and the contouring of fields, suitable for the smaller and lighter farm machinery were ploughed out and discontinued, because the modern form machinery was too large. Several other factors helped turn the public attention back to soil conservation. The environmental movement had grown, and it drew attention to soil erosion and water quality issues.

The shape of future conservation policy is uncertain, because much is being learned about soil/technology relationships and productivity gains. If one could count on continuing technical contributions to agricultural productivity at a low cost, one could discount the importance of conserving our soils. On the other hand, if productivity declines and technology are not able to match increased demands for raw food products, then it would pay to increase conservation investments.

Policy strategies for conservation have included (1) targeting particularly erosion-prone lands, (2) removing the most erosive lands, (3) mandating minimum low-cost conservation practices, (4) requiring cross-compliance of conservation practices with eligibility for participating in other programme, and (5) incentives to shift production to less erosive production. These policy options will likely not obtain forefront items in the near future, considering the economic plight of many farmers. Yet they are critical to agriculture's long-term economic survival.

Water Availability

Agricultural water use is dominated by plant production. Plants use far more water than is required for their growth and development. This inefficiency is largely a botanical consequence of stomatal apertures in the foliage through which water vapour leaves the plants. Although vapourization of water is important in cooling the leaves, loss of water vapour is not an essential function as such. Rather, it can be considered a necessary evil that accompanies absorption of atmospheric carbon dioxide through the same openings for the essential process of photosynthesis. The stomata close when the water supply is deficient, thereby decreasing transpiration but also decreasing photosynthesis.

Salinity is the most serious water quality problem in agriculture. When water evaporates from the soil or plant surfaces, the dissolved salts are left behind. These salts build up in the soil and can eventually prevent plant growth if they are not removed by leaching them out at about the same rate they are applied in irrigation water. Four independent salinity management strategies have been identified: (1) Interception of saline springs, or other point sources of saline water, and treatment by evaporation, desalting and reuse, or diversion for industrial use. (2) Conservation of the amount of water applied for irrigation. This will lessen the amount that seeps through the soil. Since seepage may displace saline groundwater or dissolve salts in the subsoil, reducing it may, in turn, reduce the salt load returned to surface or ground-water supplies. (3) Reuse of saline or sewage water for salt tolerant crops. Such use reduces the amount of brackish water needing disposal and provides a substitute irrigation water supply. (4) Change of the cropping pattern. Choosing salt-tolerant crops clearly permits use of more saline water. New plant-breeding techniques may develop more tolerant varieties of common species. However, none of these options are without cost. All have socioeconomic as well as technical consequences. Their implementation requires institutional and attitudinal changes.

Various procedures for increasing the supply of agricultural water are technically feasible. Reuse of water from municipal sources and from irrigation return flows is widely practised. Desalination of brackish water is still too expensive for practical use, and aerial seeding requires much further development. Removal of non–beneficial vegetation that has high water requirements, snow management, runoff management, and changing the landscape by directing water to limited areas are other procedures that have extensive practical use. Many possibilities exist for conserving existing supplies of water for irrigation, but there is much debate about whether the savings from individual farmer efforts will make up much towards the anticipated overall water use deficit.

As the competition for water intensifies, more attention will be devoted to devising or modifying mechanisms for allocating water. For the most part, institutional barriers have inhibited transfer of water. Perhaps the most fruitful opportunity lies in a fundamental change in the system of laws and rules allowing for voluntary exchange of water rights while protecting the interests of third parties.

In agriculture, a commonly expressed argument against relaxing institutional barriers that inhibit transfer of water rights is that higher-valued uses would take too much of the water now used for agriculture. Research does not support this view. The limited experience available suggests that market like mechanisms for transferring water work well. Further, in the future, agriculture will be held more accountable for the way it used water than it has in the past. Agriculture will have to expend more effort to justify its claims to water, particularly under scarcity conditions, and will have greater responsibility for the environmental impacts of the water it uses and returns to the system.

Current water policies that deal with some of the more prominent issues are: (1) incentives to encourage water and soil conservation practices; (2) policies to protect surface water quality by reducing allowable salt discharges into drainage water; (3) control of groundwater depletion; (4) legal and administration doctrines and rules controlling water rights; (5) the rationalization of inter basis water transfers on economic and noneconomic grounds; (6) price policies for water and water-related services should reflect these total costs (*e.g.*, why should irrigation be subsidized to produce surplus crops that other programmes pay farmers not to produce?); (7) limitations on acreage that can be irrigated with state developed water; (8) the extent to which the direct beneficiaries of publicly developed water should be required to bear its development and operating costs; (9) the extent to which competitive interest rates should be charged for funding public water development projects; and (10) the impact on water use of government policies that are seemingly unrelated to water.

Energy

Generally, fossil energy has been used to perform three basic tasks; to replace humans and draught animals in the performing of farm work, to increase yields, and to preserve and protect food once it is grown. Three important concerns prevail; whether the energy will be available, what it will cost, and whether some alternative forms of energy might be used to accomplish these tasks.

The first two tasks fossil energy performs in food production take place on the farm. People have focused on this energy use because it represents the starting link of the food chain and because it has been easier to measure than the more diverse uses of energy once the food has left the farm. Agricultural production directly consumes 2-3% of the total fossil energy used in the USA. Food processing uses 5-7%, home preparation 4-6%, retailing and wholesaling 2-3%, and transportation uses 1-2%, making a total of 14-21% of the nation's fossil energy now being used in the food system.

Dependable fuel availability without cutoffs or shortages is critical to the food system. Producing food is a biological process. Once set in motion, resources committed to this process cannot be reclaimed. In addition, the food product is subject to spoilage, and preventing spoilage requires energy.

The energy required to do farm work is largely supplied by liquid petroleum fuels supplemented by natural gas and electricity. Any one of these can be crucial to a particular type of farming operation. Ploughing or harvesting might be impossible without diesel or petrol for the tractor; irrigation pumps might not be stalled without sufficient funds.

Liquid petroleum fuels are used to do farm work. Sources of petroleum are geographically diverse, and there are some national stocks in the strategic petroleum reserve, and short-term shortages would be met to some extent by drawing down these stocks. The question is how these stocks will be allocated and how quickly and effectively they can be obtained by end users? If the strategic reserve is to be mobilized quickly in an oil cutoff, pumps, pipelines, and refineries have to be in place and dedicated to such emergency use. The planning for and management of the strategic petroleum reserve are critical for farmers, and the answer will help determine their ability to meet absolute shortages of liquid petroleum fuels. However, there has been little concern about the reserve and its management in the 1980s, so it may not be fully able to meet an emergency situation.

Second, the value of energy used in the overall food and fibre sector has been only a small fraction of the value of the total product sold by that sector. The value of the direct energy input as a percentage of the value of the total product would be 5-7% in food production, about 2-30% in food processing, about 5% in food marketing, 8-9% in the preparation of food at home, and probably

15-16% in food preparation away from home. Specific products or activities are much more important in energy-monetary terms, and new technology is being developed and applied to cut energy costs. A great deal of effort is being made to cut energy costs in food preparation away from home.

Many of the alternative energy sources now being considered are based on biomass production. Shortages of high-quality liquid fuels have been of primary concern, and numerous agricultural experiments have been undertaken in the search for substitute fuels. Many countries have large supplies of relatively low cost energy from coal. Most operations that require energy, with the exception of mobile power requirements, which need liquid fuels, could be accomplished by using coal as a basic source of energy. In this sense, coal and the cost of technology for burning or liquefying it place a cap on the price to be charged for other fuels or energy forms sold to perform these tasks.

Food and fibre crops appear to have an edge over energy crops in the competition for land, and this will continue unless there is a large relative price shift between food and the agricultural products used for energy. We do not have a significant unused land base suitable for intensive cultivation of energy or food crops. Unused land is often unused for very good reasons. There are times when we have excess food production capacity in the form of excess cropland, but this has tended to be needed on a cyclical basis from time to time. Its cyclical need in food production limits its commitment for the time necessary to produce fuel and recover the cost of a capital-intensive processing plant. If we had adequate economic incentives, we could probably convert 75-100 million hectares from nonfood and fibre to biomass production. However, growth in the demand for food might result in the conversion of half of this row-crop acreage back to food production in 20-30 years.

If one looks at the energy cost of high-pressure irrigation or the alternative use for electricity generation of gravity-fed systems, it does not make sense to irrigate land to grow energy crops. Water is likely to be increasingly scarce for agriculture and other uses, so it makes little sense to plan future biomass production based on the necessity to irrigate. This places definite limits on the amount and location of land available for biomass energy production, limits that are not always recognized. Ethanol is already being produced from grain.

There are two approaches to the future. If we face absolute shortages of critical fossil energy, we will face revolutionary approaches to meeting these needs. An interruption of foreign petroleum supplies would result initially in allocation (by price or by government) and a scrambling to produce alternatives. The longer-term substitute would probably be coal liquids, but in the short term, modest amounts of liquid fuels from biomass could be the need of time, form and place. That is why something like the strategic petroleum reserve is especially critical to agriculture.

Policy Planning

Planning is basic to a systematic consideration of the future. Government planning itself is not action; it does not directly alter access to resources. It affects policy by providing information on the social cost/price off the options available, including the option of no action. Planning also identifies the logic behind any rule change that alters resource use, including the preferences of those who do the planning. Issues evolving from the use of natural resources can be highly complex and controversial, and actions taken to solve one problem may create a dozen others. The most important future issues are likely to be the following:

Agricultural Pollution

Soil erosion may only mean that soil is moved from one field to another. However, if that sediment carries various nutrients or chemicals, it can cause problems for other water users. Environmental degradation through excessive use of agrochemical draws attention of planners and scientists. Educating farmers to understanding the off-site effects of erosion to the outright prohibition of farming practices that permit unacceptable levels of runoff. Thus agricultural pollution from various sources will be a very important issue in years to come.

Wetland Protection

Wetlands or swamps may cause problems to farmers, but they are highly valued by many environmental groups. In an effort to protect natural wetlands, recent policies have discouraged or even prohibited farmers and others from draining them.

Biomass Energy

Animal and plant organic matter can be an important source of energy production. Farmers have been among the first to use biomass as energy for their own needs. Serious attention is being given to the intensive production of energy crops. The issue of food or fuel, with all its related questions of feeding the starving populations of the world versus providing adequate energy for the energy-intensive life-style in the developed countries, will become increasingly important.

Farmland as a Waste Processor

The enormous natural cleansing capacity of soil has been marshaled by communities and industries in all the developed countries. Waste water and sludge from municipal treatment plants provide sources of moisture, nutrients, and organic matter that farmers may value. However, the waste may also contain sufficient toxic substances to be unacceptable. Health departments have been cautious about permitting applications of human or industrial waste on land used to produce food for direct human consumption. But within limits of timing and volume, farmland can be a natural and productive waste-treatment system that provides an important resource service in the future.

Noise and Nuisance

Agriculture is frequently a noisy and smelly occupation, tolerated by some people but not others. Reasonable objections are raised to some farm practices. As a result, various nuisance ordinances and anti–farming regulations have been passed in communities in several countries of the world. People have sought relief from what they deemed to be unacceptable interference with their enjoyment of the combination of natural resources and a rural life-style. The policy issue is that in order to remain competitive and economically viable, agricultural producers need certain kinds of freedom to exercise normal farming practices.

Labour

Farm labour policies have been aimed at improving the lot of hired workers. State protective legislation has gradually been extended to farm workers including minimum-wage legislation, worker's compensation, unemployment insurance, and health and safety regulations. However, much of this legislation covers only farms with a minimum number of workers, so it may not apply to the typical family farm.

Terms of employment conditions of work, and wages/hours/ benefits will dominate collective bargaining issues in the food and fibre system. Labour contractors will continue to be important where seasonal employment is high, such as in fruit and vegetable farm areas. Employer/employee relationships will continue to change at the farm level, even for owner-operators who want to hire temporary casual labour. Definitions of work, recruitment, orientation, motivation, and performance evaluation will gain recognized importance on the larger commercial farms.

Labour policy throughout the farm and fibre system in the next 15-20 years, while aimed at improving the lot of workers, will accept that a reasonably defined set of expectations from both management and labour will help the situation. It will also challenge both labour and management to understand that the next 20 years will not be like the past 20 years. Business will not be as usual anywhere in the system. Therefore inventiveness to find mutually beneficial ways to settle differences with minimum cost to each will be a key issue for both parties.

Management

Management in all parts of the food and fibre system will need to become more knowledgeable and sophisticated about new forms of business technology. As production concentrates in fewer managerial hands throughout the system, managers will increase their efforts (1) to substitute capital for labour wherever possible and profitable, (2) to enhance labour productivity through capital investment, and (3) to simultaneously work hard on developing good relations with their workers.

Acceptance of the industrialization concept about agriculture, along with its growing concentration and competitive nature, will eventually force a different set of farm and food policies from the government. There will be adjustment problems to those policy changes. New issues will evolve that cannot be identified now. However, successful managers of the future will devise anticipatory strategies, use more financial and market sensitivity analysis, and become more knowledgeable about systems of farming and integrating production/marketing decisions than they are currently.

Transportation

Transportation costs are subjected to many market pressures (back-haul capacity, energy, labour, the value of the commodity transported, insurance, etc.) as well as institutional pressures (locked-in labour contracts; different unloading, handling, and loading contracts at various points; differences in country-to-country shipment costs; and the effects of truck and rail deregulation). Increasing competition for markets will force increased managerial pressures for cost cutting to include transportation.

WILDLIFE AND AGRICULTURAL ISSUES

Wildlife helps agriculture by providing the genetic mass that guards against too homogeneous a biological base which, in turn, lessens the chance of a catastrophic attack on our raw food supplies. Genetic diversity is, indeed, the preservation and future of all life forms. Research values for many species, even seemingly insignificant ones, can provide biologically important information. Many wild plants are important sources of medicine.

Non monetory values are associated with agriculture and wildlife, values that are educational, aesthetic, psychological, and philosophical. These values have led to the enactment of the Endangered Species Act and the National Environmental Policy Act in many countries, which have a great impact on the farmers. The underlying philosophical tenet for these laws is that the future of everyone depends on the condition of the entire earth. The policy approaches to preserving and extending habitats of wildlife have gained momentum in the last 10 years and likely to increase in the next decade.

While governments have done a great deal to influence wildlife/agricultural ties, a great proportion of wildland and wildlife preservation has been done by individuals and private organizations without recourse to public coffers. Indeed, many private sector efforts have encountered a seemingly inflexible public bureaucracy incapable of responding to rapidly changing political pressures, the evidence of credible research, or the calls of common sense. The interdependence between agricultural production and wildlife species is complex. Unabated habitat losses will lead to further species losses, thereby retarding agriculture's ability to adjust to future environmental changes.

AIR ISSUES

Agricultural production results in some air quality problems. Examples are dust, odours, and pesticide drift. However, when compared to the electric power industry, paper mills, and automobiles, agriculture is a minor source of air pollutants. Of greater concern for agriculture are the effects that air pollution from other sources has on agricultural productivity. Crop damages have been well documented near urban and industrial areas. Furthermore, air pollutants can travel hundreds of kilometers to damage crops far distant from urban and industrial sources.

Ozone is currently receiving the attention from scientists as a threat to agriculture. This air pollutant is widely distributed in rural areas. Ozone is thought to be responsible for up to 90% of the national crop losses caused by air pollutants. These pollutants originate primarily from the combustion of fossil fuels used for electric power generation and transportation. Estimates presented by the Office of Technology Assessment (USA) indicate that reducing ozone concentrations to natural background levels would have increased maize yields nationwide by 3%, wheat by 8%, soybean by 17%, and groundnuts by 30% in 1978. Losses must have intensified to a much higher level since then all over the world.

Human activities can cause climate changes and are becoming an emerging environmental issue with far-reaching implications. The climate of an area is typically defined as the long-term average weather of an area. Deviations from the average, are not climate change but variations in climate. However, activities change the character of the earth's surface through urbanization, changes in agricultural land use, and waste heat generation.

One climatic issue that has achieved particular prominence is the greenhouse effect, resulting from the growing concentration of carbon dioxide in the atmosphere. Half of this increase is believed to have occurred within the last 20 years and is generally attributed to industrial growth and the tremendous growth in fossil fuel combustion during this time. Scientists state that rising atmospheric concentration of carbon dioxide has initiated a global warming trend (greenhouse effect) and disrupting global weather patterns.

METEOROLOGY AND CLIMATE CHANGE

High-production technology is strongly linked to good weather conditions. The technical capability of estimating short-and long-term changes in climate is essential for growers in attempting to manage risks due to weather. Drought, flood, hail, and storms, extremely low and high temperatures- all can be detrimental to agricultural producers if not planned for and protected against. Automatic surface weather stations, satellites, and special electronic sensing devices accumulate vast amounts of information that can be processed by computer to forecast weather and climate changes. Many organizations (national and international) contribute to a common data pool for these forecasts. Most information focuses on weekly, monthly, or seasonal values of important variables such as precipitation and temperature. Of greater importance are the deviations from normal values, used to recognize anomalous weather events.

Humans have increased the turbidity of the lower atmosphere by adding industrial waste to the atmosphere. The net effect of these increases in turbidity is a cooling of the earth. Certain atmospheric gases selectively absorb the infrared radiation emitted from the surface of the earth. These gases reemit infrared radiation towards the earth's surface and top of the atmosphere. The net effect is an equilibrium balance of radiation, which helps to determine the earth's surface temperature and the temperature of the lower atmosphere. A change in the concentration of the gas or gases involved in absorbing infrared radiation will produce a new equilibrium temperature. Increased concentration of these gases will cause a warming at the earth's surface, and vice versa. Good absorbers of infrared radiation include water vapour, carbon dioxide, and oxides of nitrogen. Carbon dioxide concentrations have been increasing and are expected to continue to increase in the foreseeable future. By the year 2025 the level of CO_2 is expected to be two and a half times the current concentration. This could increase the mean temperature of the earth and lower atmospheric temperature by 2-3°C.

Climate and weather may be the variables that cause agriculture to make a necessary reality check. Both of these variables have controlled raw food production since time immemorial. Man continually tries to improve them by introducing new technology in hopes of controlling his destiny. What the world is learning, albeit slowly, is that hi-tech agriculture, as it is now popularly envisioned, may create more problems than it solves, that it leaves more questions than answers. It is to this point that future technological challenges are directed. We must all come to grips with the age-old conundrum of how to provide sustenance for the present without crippling the future. Moving to a new scientific frontier will not be easy- economically or politically. Yet it is becoming abundantly clear and that is what is needed for agriculture.

BIOTECHNOLOGY

One growing area of research is biotechnology. In agriculture, biotechnology possesses an, as yet, undefined potential for increased control and technique for processing and manufacturing new plant and animal products. For example, scientists are working on using recombinant DNA techniques and protoplast fusion to provide the capability to transfer genes between plants and between animals. It may even be possible to bring together populations of genes for which natural barriers to sexual reproduction currently exist. Engineering plants with such characteristics as nitrogen-fixing abilities, resistance to bacteria or viruses, salt and drought tolerance, and perenniality are also close. Other future possibilities include pesticides and herbicides that can be used to prevent specific diseases or pests, vaccines, plant and animal growth hormones, and fertilizer replacements. A final use may be in industrial processes (e.g., fermentation) to produce foods formerly obtained from field agriculture).

Biotechnology is already making large inroads into animal research. One example of a newly developed biotechnology is bovine growth hormone (BGH). Depending on the stage of lactation, 44 mg of BGH injected into a dairy cow daily can be expected to result in a 15-40% increase in milk production regardless of the cow's normal production level. Another example is the development of embryo transfer. It is now possible to transplant live embryos of high-quality dairy and beef cattle to recipient cows in order to gestate and produce genetically superior cattle. Through cryo-biology, these embryos may be frozen, stored and transported for use at any time and any place in much the same way that semen has been stored and transplanted for years.

New types of plant life than can, for example, biologically fix nitrogen, could have a tremendous effect on the demand for natural or man-made nitrogenous fertilizers. Plant life (including bacteria) that is genetically adapted to resist selected bacteria or viruses; to tolerate salt, drought, and freezing; or to retain immunity to pesticides may have dramatic effects on the cost of inputs and location of production. Plant life that is genetically engineered to be perennial or to have increased photosynthetic efficiency or increased hybrid vigour may affect the magnitude of production. The possibilities of producing energy-creating crops at levels that make them cost-effective may further reduce total farm production costs.

Biotechnology in general, and genetic engineering in particular, may be suffering from too much attention about current and potential research. This is due in part to a general confusion about what the large capital investments in biotechnology are for and in part to the use of terms like biotechnology and genetic engineering. These confusions are valid and need to be cleared up. Biotechnology is not a discipline but a field of activity that applies scientific and engineering principles to the processing of materials by biological agents to provide goods and services. Current examples include sweeteners and protein synthesized by bacteria as well as field production of food, fibre, and industrial feed stocks. Genetic engineering is not biotechnology but is the scientific development of techniques for genetic manipulation that will have a substantial impact on biotechnology. What can realistically be expected from these research areas in the future and what effect will these changes have on the farm and food system is simply not yet known.

Major technical constraints currently limit the full realization of biotechnology. For example, scientists do not yet fully understand how to identify genetic functions and link them to agronomic traits because of the complex interactions between the genetic materials of various crops. Success with simple microorganisms does not imply equal success with higher-order plants and animals. What appears to be possible in theory is proving extremely difficult in practice.

SUSTAINABLE AGRICULTURE

Existing agriculture has sustained large population increases, improvements in human health, and escalating standards of living for many people. This productive agriculture has included a wide range of petrochemicals to those based solely on organic techniques- the system chosen depending on culture, availability of resources, technology, and economics. For all its achievements in sustaining life to the present time, however, there is increasing concern that some of the practices in present-day agricultural systems are damaging the resources on which agriculture is based. A worsening of soil and water quality, a recognition that a large percentage of modern farming is dependent on some inputs that are potentially hazardous and nonrenewable, and concerns about food safety and human health, all raise questions about whether some current agricultural systems can or should continue.

The benefits of a flourishing agriculture have in some cases generated high costs in natural resources, in energy, in environmental quality, and at the present time, paradoxically, in reduced profitability to farmers. There is an increasing awareness that the sustainability of agriculture and its many contributions will require alteration of some existing practices and systems.

As the term sustainability is currently used, it is an assertion about the future. It is a judgment about whether or not certain agricultural practices will conserve natural resources, improve economic returns and continue to enhance productivity and human lives in the long term. As increased knowledge and experience improve judgments abut the use of resources, sustainability can be understood as a process through which farming and resource management systems are developed to permit a regenerative agriculture. Viewed as a process, sustainability becomes a dynamic task, enabling and encouraging greater awareness of the complexities and interactions among physical, biological, political, cultural, and economic factors. This awareness calls for an increasingly sophisticated approach to the analysis of resources, to the understanding of environmental issues, and to the management of highly complex farming systems. It calls for a long-term dedication and commitment, both from those who till the land and from those who conduct research and educational programmes to improve and sustain agriculture.

Sustainable programmes and practices are designed to decrease dependence upon costly, nonrenewable production inputs and to build the soil base to ensure inherent long-run productivity including protection from plant and animal diseases and other pests of all kinds. They are designed to generate increased awareness of relationships between plants, soil, air, and water under differing ecological conditions of climate and biological stress. Sustainable activities also direct attention to the relationships between livestock and fish production with their dependence on the vegetative productivity of the world's land and waters. The ultimate goal is to improve the quality of human life based on our ability to manage, adapt, and modify natural resource systems. Included in this goal is the capacity to help provide adequate food and fibre throughout the world by drawing on both agriculture's present capabilities and its future potential.

The current economic problems in agriculture, and the immediate concerns about the stress to which we have put our natural and human resources, should serve as an incentive for long-range research and information dissemination programmes. Universities and institutions hold no monopoly on the creative work done in agriculture, yet they do offer a proven system for developing and extending information applicable to concerns about the future of agriculture. People who study and practice agriculture together quickly discover there are no once-and-for-all answers. It is precisely the quest of long-term answers, however, that will help achieve a truly sustainable agriculture for us all.

The physical resources of air, land, and water will undergo policy issue debates as chemicals, energy, labour, and engineering are applied to them. Biological resource policy issues relating to animals, plants, pests, food, and nutrition will also arise as people use the products of combined resources. Ultimately, decisions will be made within a cultural context based on economic and/or political reasoning.

It is clear that unleashed technology based on petrochemicals can reap abundance in the short run at considerable economic cost. Long-run costs are also borne by these same resources and by future generations of mankind. We have farmed in an era of food scarcity and inexpensive production inputs. Future farm business cannot be as usual- cannot because much of today's farming is not physically, economically, or socially viable in the long run.

What the answers will be is not clear. Much research needs to be done. However, with increasing regulation about the use of chemicals in agriculture, more focus will be put on how to enhance the inherent productive qualities of soil through different farming systems and cropping patterns. Sustainable, or regenerative, agriculture will always have to confront the urgencies of time as it also faces the question of how to help feed people profitably.

RESEARCH AND DEVELOPMENT NEEDS UNDER ORGANIC FARMING

World Agriculture

The impact of structural changes in agriculture in future on the prospects of organic farming may involve four key issues. The first issue is the size of the organic farms. In USA and other developed countries, large scale farms exist and for conversion of conventional farmers to organic farming may prove to be somewhat risky.

Organic farming revolves around minimum use of synthetically produced inputs such as inorganic fertilizers and pesticides. Thus, small scale farmers will find it more convenient to take up organic farming with a lesser degree of risk attached than will large scale farmers. Land holding of farmers in India and some other developing countries will easily suit adoption of organic farming practices. The extent of organic farming may be gauged in terms of number of organic farmers who practice organic farming methods and the area covered under organic farming systems. From the experience of several countries, where farmers have switched over to organic farming from high-tech farming, it can be gauged that the number of farmers rather than the area covered is more important.

Organic farming depends heavily on easy access to supplies of non-chemical inputs such as organic manures and other wastes. Thus, the adoption rate shall directly depend on availability of the derived organic inputs backed up by technology or package of practices.

The fourth issue regarding conversion of conventional farmers to organic farming structure is the role of energy price. Because conventional farmers are accustomed to energy intensive practices, they will have to spend more in future to produce in view of shrinking fossil fuel reserves and subsequent price escalation. Increase in prices of farm produced commodities will play an important role in influencing a conventional farmer towards adopting organic farming. Rapid increase in prices of farm commodities would make the conversion relatively unlikely while stability in prices shall create conditions more favourable for adoption of organic practices.

The available information on organic farming is very meagre and some of the research undertaken during the era before the use of chemical fertilizers and pesticides has relevance to today's organic farming. Much progress has been made in all fields of agriculture during the past four decades as far as the conventional high-tech agriculture. However, this research has very little appreciation with regard to organic farming and specially those concerning the sustainable systems. Moreover, we must realize that the future progress of the organic farming systems will largely depend on generation of new technology suitable to a particular agroclimatic condition under the present structural set up.

The development of sustainable farming systems will require inter disciplinary approach to research on resource conservation, reduced tillage, pest management, crop rotations, improved crop varieties etc. The extension functionaries at different levels shall also need to be attuned to the change in farming concept *i.e.* from energy intensive agriculture to organic agriculture.

Youngberg and Buhel (1984) had suggested the following priority research areas of organic farming: Although each of these research areas are important, the ultimate scope will depend on the location, timing of the agenda and interest of the researcher and the institution. By and large, these cover most of considerations envisaged under organic farming.

- **The ideology and politics of organic agriculture**. What is the present nature of organic philosophy and politics? Is it changing? How? Why? Here, studies of the conventional, general and commodity farm organizations should be examined with respect to the ideological and political overlap and conflict between their policy objectives and those of organic farmers.
- **The sociology of agricultural science**. The role of professional socialization and the status and reward system of professional organizations in determining the overall agricultural research agenda. What are the implications of this system for organic farming research?
- **The changing nature of alternative agriculture as a symbolic phenomenon**. What meanings do individual scientists,

politicians, administrators, conventional farmers, and lay citizens attach to such terms as organic, biological, natural, and ecological farming? What do such terms mean to them? What are their perceptions of these modes of agriculture? What are the implications of these meanings and perceptions for the adoption of alternative systems? To what extent and in what ways does the symbolism of alternative agriculture either facilitate or impede the adoption of such systems?

- **The implications of production declines under organic farming systems**. What are the likely consequences of production declines and a changed product mix which would occur with the widespread adoption of organic farming? What would result? In particular, what effects would these results have upon domestic consumers, export trade, Third World hunger, input and product prices, distributional price increases, land and asset values, etc.?
- **Public policies and their relationship to organic farming**. Agricultural commodity, marketing, trade, taxation, credit, conservation, research, regulatory and incentive policy should be examined in light of its relationship to, and impact upon, the development of organic farming systems. State and local governmental policy should also be included in these studies.
- **The demography of organic agriculture**. Major knowledge gaps remain with regard to the numbers and sizes or organic farms as well as the ideological and socioeconomic character of organic farmers. How many organic farmers are there in the country? How do they farm? What role does the farm family play in the success of an organic farm? In particular, policymakers and would-be organic farmers would benefit greatly from systematic and comprehensive socioeconomic studies of organic farming production systems on a variety of sizes and types of organic farms. Until reliable and scientifically acceptable information is obtained with regard to the micro-socioeconomic character of organic farming, it is virtually impossible to assess the macro-economic and policy-related effects of the wider scale adoption of organic farming systems.
- **Recycling and land application of urban organic wastes**. What are the local politics of urban waste management?

What are the consumer attitudes regarding various recycling methods and strategies?

- **The modeling of energy price increases and supply interruptions, particularly with regard to constraints on farmer decision-making and macro-economic effects.** What affects would various developments in these areas have upon food prices, food security, domestic and international food supplies, exports and other facets of international relations?
- **The scientific literature of organic farming and the character and status of educational materials related to organic farming.** Inventory, analyze, and assess existing research reports and professional publications, extension materials, and other sources of information with regard to their relevance and applicability to the informational and educational needs of contemporary organic farmers. In addition to the practical farmer benefits to be derived from such an inventory, it could also serve as a guide to current researchers as they plan future long-range research projects and programme in this area.
- **Technology transfer.** Given the complexity and holistic nature of organic farming systems, are there unique problems associated with the information and technology transfer functions, either from the perspective of the provider or recipient of information?
- **Agricultural courses, curricula, and text books.** How do existing agricultural curricula and text books present materials related to organic farming? Do such sources accurately reflect and portray the character of contemporary organic agriculture?
- **The machinery needs of organic farmers.** Do organic farmers, especially small-scale ones, have unique machinery needs? What are the socio-economic implications of these needs with respect to adoption capabilities and technology transfer, generally? If such machinery were available would it be economically feasible for small farmers to adopt it? Would special incentives and educational programmes be required? How would these needs vary from one farming region and type of organic farm to another? What are the international implications in this area?

Indian Agriculture

India is not in a position at present to completely do away with the use of synthetic agrochemicals, especially the inorganic fertilizers, in view of the large demand of the increasing population for food commodities. Systematic phasing out of the use of agrochemicals and synthetic fertilizers and increasing the use of organic manures, reduced tillage biofertilizers etc. may prove beneficial in this direction. At present the country lacks a systematic research and development programme in respect of sustaining agricultural systems though organic farming practices. The following priority areas for research and development under organic farming are recommended.

- Research projects for formulating organic farming practices should be framed by ICAR and other agencies.
- Establishment of a national centre for organic farming will be useful in undertaking and coordinating basic research on organic farming systems.
- Development of pesticides of plant origin (*e.g.*, neem) and use of bioagents especially under Integrated Pest Management systems need to be promoted.
- Linkages between the Ministry of Agriculture, Government of India, ICAR, State Departments of Agriculture and the SAU's need strengthening in respect of developing organic farming systems.
- SAU's and ICAR institutes should restructure their course curriculum by including courses on organic farming.
- Incentives for production of good quality organic manure, biopesticides, biofertilizers and green manuring crops may be considered.
- Encourage visits of farmers to model organic farms. Organise seminars, workshops, conferences and lectures to create awareness regarding the important components of organic farming.
- The indiscriminate use of inorganic fertilizers and pesticides need to be discouraged.
- Steps are needed to avoid hazardous chemical residues in

feeds, fodder, food products and milk.

- Develop marketing infrastructure for organically produced commodities.
- Financial institution can come forward to provide appropriate support.

The technological and economic feasibility of organic farming will certainly influence its large scale adoption and promotion. Coupled with this, the need for alternate farming systems in view of emerging environmental problems and energy constraints will have a bearing on the agricultural structure in general, and adoption of organic farming practices, in particular. However, awareness about degradation of earth's natural resources at an alarming rate and man's quest to survive and keep his planet green and healthy should go in favour of world wide adoption organic farming systems.

References

Ahmed, S., Grainge, M., Hylin, J.W., Mitchel, W.L. and Litsinger, J.A. 1984. Some promising plant species for use as pest control agents under traditional farming systems. In: Proc. 2nd Int. Neem Conf., Rauischholhausen, pp 565-580.

Alexander, M. 1977. Introduction to soil microbiology. John Wiley & Sons, New York.

Alien, M.F., Moore, T.S. and Christensen, M. 1982. Canad. J. Biol. 60: 468-477.

Anonymous. 1992. Fertilizers account for 75% of Asian cereal growth. Fert. Agric. March, 1992: l.

Arar, A. 1988. Background to treatment and use of sewage effluent. Ch. 2, Treatment and Use of Sewage Effluent for Irrigation, M.B. Pescod and A. Arar (eds). Butterworths, Sevenoaks, Kent.

Arnason, J.T., Philogene, B.J.R. and Morland, P. 1989. Insecticides of Plant Origin. American Chemical Society, Washington, DC.

Arora, C.L., Nayyar, V.K. and Randhawa, N.S. 1975. Note on secondary and microelement contents of fertilizers and manures. Indian J. Agric. Sci. 45: 80-85.

Asano, T., Smith, R.G. and Tchobanoglous, G. 1985. Municipal wastewater: Treatment and reclaimed water characteristics. Irrigation with Reclaimed Municipal Wastewater- A Guidance Manual, G.S. Pettygrove and T. Asano (eds.). Lewis Publishers Inc., Chelsea, Mississipi.

Asano, T., Smith, R.G., Tchobanoglous, G. 1985. Municipal wastewater: treatment and reclaimed water characteristics. In: Pettygrove, G.S., Asano, T. (eds). Irrigation with reclaimed municipal wastewater- a guidance manual. Lewis, Chelsea, pp 2.1-2.26.

Ayers, R.S., Westcot, D.W. 1985. Water quality for agriculture, Rev edn.

FAO Irrig. Drain Pap 29. Rome. 174 pp.

Badji, Moussa and Sene, Modou. 1992. Sustainable Land Use Systems Research in West Africa Sahel Proc. International Workshop on Sustainable Land Use Systems Res., New Delhi, India, Feb. 12-16, 1990.

Baier, D.C., Fryer, W.B. 1973. Undesirable plant responses with sewage irrigation. J. Irrig. Drain Div. ASCE 99: 133-141.

Bala, S.K. and Sukul, N.C. 1987. Systemic nematicidal effect of eugenol. Nematropica 17:219-222.

Batie, S. 1989. Sustainable Development: Challenges to the Profession of Agricultural Economics. Amer. J. Agri. Econ.-71: 1083-1101.

Baujard, P., Chabrier, C., Martini, B., Meunier, L., Pariselle, A. and Sarr, E. 1989. Comparison of seven nematicides and studies on the use of dibromochloropropane in groundnut culture in the Sahelian area of Senegal. Rev. de Nematol. 12: 293-299.

Bay-Peterson, J. 1986. New technology for low-income farmers in developing countries. Outlook Agric. 15: 110-114.

Bezdieck, D.F., Power, J.F,, Keeney, D.R., Wright, M.J. (ed). 1984. Organic Farming: Current Technology and its Role in a Sustainable Agriculture, Am. Soc. of Agron., Crop Sci. Soc. of Am., and Soil Sci. Soc. of Am. Spec Pub. No. 46, Am. Soc. of Agron, Madison.

Bhardwaj, K.K. and Gaur, A.C. 1985. Recycling of Organic Wastes. ICAR, New Delhi.

Bhardwaj, K.K.R. and Gaur, A.C. 1985. Recycling of organic wastes. ICAR, New Delhi, pp 104.

Bijay-Singh and Dhaliwal, G.S. 1993. Eutrophication: A threat to wetlands. In: G.S. Dhaliwal, B.S. Hansra and S.S. Ladhar, (eds). Wetlands: Their Conservation and Management Punjab Agricultural University, Ludhiana, pp 95-108.

Bijay-Singh and Sekhon, G.S. 1977b. Impact of fertilizer use on environmental pollution in Punjab- Present status and future projections. Pert. News 23(2): 7-11.

Bijay-Singh and Sekhon, G.S. 1979. Nitrate pollution of ground water from farm use of nitrogen fertilizers- a review Agric. Environ. 4: 207-22S.

Bijay-Singh, Biswas, C.R. and Sekhon, G.S. 1978. A rational approach for optimizing application rates of fertilizer nitrogen to reduce nitrate pollution of natural waters. Agric. Environ. 4: 57-64.

Bingham, F.T. Page, A.L., Mahler, RJ., Ganje, T.J. 1976. Yield and cadmium accumulation of forage species in relation to cadmium content of sludge-amended soil. J. Environ. Qual. 5: 57-60.

Bingham, F.T. Page, A.L., Mahler, R.J., Ganje, TJ. 1975. Growth and cadmium accumulation of plants grown on a soil treated with a cadmium- enriched sewage sludge. J Environ. QuaL4: 207-211.

Black, C.A. 1968. Soil-plant relationships, 2nd edn. John Wiley & Sons, New York.

Bouwman, A.F. 1990. Soils and Greenhouse Effect. Wiley, Chichester.

Brezonik, P.L. 1978. Nitrates: An Environmental Assessment. National Acad. Sci., Washington, DC.

Canway, G.R. 1986. Agroecosystem analysis for research and development, Winrock International, Bangkok.

Chaney, R.L. 1989. Toxic element accumulation in soils and crops: protecting soil fertility and agricultural food-chains. In: Bar-Yosef, B., Barrow, N.J., Goldshmid, J. (eds). Inorganic contamination of the vadose zone. Springer Ecol Stud 74:140-158.

Chang, A.C., Page, A.L. 1985. Fate of wastewater constituents in soil and groundwater: trace organics. In: Pettygrove, G.S., Asano, T. (eds). Irrigation with reclaimed municipal wastewater- a guidance manual. Lewis, Chelsea, pp 15.1-15.20.

Chatterjee, A., Sukul, N.C., Laskar, S. and Ghoshmajumdar, S. 1982. Nematicidal principles from two species of Lamiaceae. J. Nematol. 14: 118- 120.

Cooper, P.P., Hobson, J.A. and Jones, S. 1988. Sewage treatment by reed bed systems: The present situation in the UK. Water Research Centre, Stevenage.

Crutzen, P.J. 1981. Atmospheric chemical processes of the oxides of nitrogen, including nitrous oxide. In: C.C. Delwiche, (ed). Denitrification, Nitrification and Atmospheric Nitrous Oxide. Wiley, New York, pp 17-44.

Dag, J., Feigin, A., Giskin, M., Golan, S,, Jerome, G., Davidov, S., Keinan. M. 1984. Response of cotton to effluent irrigation in a grumusol soil in the Yizre'el Valley; 1976-1982 experiments. 1: Cotton yield. Agric. Res. Org. Bet. Dagan, Isr. Pamphi 228:47 pp.

Dahama, A, K., Sinha, M.N. and Rai, R.K. 1984. Relative Efficiency of Grain Legumes in Utilizing Soil Phosphorus. Ann. agric. Res. 5 (1-2): 156-160.

Dahama, A. K., and Sinha, M.N. 1985. A study on NP concentration and their uptake. Indian J. Agron. 30 (4): 422-428.

Dahama, A. K., Sinha, M.N. and Rai, R.K. 1984. Studies on Uptake and Utilization of Phosphorus by some Grain Legumes. J. Nuclear Agric. BioL Vol. 13, pp 62-63.

Dahama, A.K. and Sinha, M. N. 1985. Effect of Kharif Grain Legumes and Residual Effect of P Applied to them on Succeeding Wheat. Ann. agric. Res. 6(1): 55-59.

Dahama, A.K. and Sinha, M.N. 1983. Tracer Studies on the Utilization of P by Wheat Grown After Kharif Grain Legumes. Ann. agric. Res. 4 (1-2): 61- 68.

Datta N.P. and Srivastava. S.C. 1963. J. Ind. Soc. Soil. Sci. II: 189.

Dennis, P. Garrity, and Percy E. Sajise. 1992. Sustainable Land Use Systems Research in Southeast Asia: A Regional Assessment. Proc. International Workshop on Sustainable Land Use Systems Res., New Delhi, India, Feb. 12- 16.1990.

Dhaliwal, G.S. and Arora, R. 1993. Changing status of insect pests and their management strategies. In: K.S. Gill, G.S. Dhaliwal and B.S. Hansra (eds). Changing Scenario of Indian Agriculture. Commonwealth Publishers, New Delhi, pp 98-145.

Dhaliwal, G.S. and Singh, B. (eds). 1993. Pesticides: Their Ecological Impact in Developing Countries. Commonwealth Publishers, New Delhi.

Drefus, B.L. and Dammergues, Y.R. 1981. F.E.M.S. Microbial Lett. 10:313.

Drupka, Stanislaw. 1992. Sustainable Land Use Systems Research in Eastern Europe. Proc. International Workshop on Sustainable Land Use Systems Res., New Delhi, India, Feb. 12-16, 1990.

Eraser, P. and Chilvers, C. 1981. Health aspects of nitrate in drinking water. Sd Total Envir. 18:103-116.

FAO (Food and Agriculture Organisation). 1985. FAO Production Year Book, Rome.

FAO. 1988. Irrigation practice and water management. L.D. Doneen and D.W. Westcot Irrigation and Drainage Paper I, Rev. 1. FAO, Rome. 71 p.

Feigin, A Biclorai, H., Shalhevet, L, Kipnis, T, Dag, J. 1979. The effectiveness of some crops in removing minerals from soils irrigated with sewage effluent. Prog. Water Technol. 11:151-162.

Feigin, A., Bidomi, H. Dag, Y., Kipnis, T, Giskin. M. 1978. The nitrogen factor in the management of effluent-irrigated soils. Soil Sci. 125:248-254.

Feigin, A., Vaisman, 1., Bieiorai, H. 1984. Drip irrigation of cotton with treated municipal effluents. 2: Nutrient availability in soil. J. Environ. Qual 13:234-238

Follett, R.F. 1992. Nutrient related pollution issues for managing nutrients. In: M.S. Bajwa, N.S. Pasricha. P.S. Sidhu, M.R. Chaudhary, DX. Bembi and V. Beri, (eds). Management for Sustained Productivity. Department of Soils, PAU, Ludhiana, pp 227-243.

Francis, C.A. Hardwood, R.R. and Parr, J.F. 1986. The potential for regenerative agriculture in the developing world. Am. J. Alt. Agr. 1: 65-74.

Gaur, A.C. 1969. Agrochimica, 14: 62.

Gaur, A.C. 1978. Proc. India/FAO/Norad Seminar on the development of the complementary use of mineral fertilizers and organic materials in India. Ministry of Agriculture, N. Delhi, pp 109.

Gaur, A.C. and Mathur, R.S. 1979. Indian I Agron. 24: 449.

Gaur, A.C. and Singh, G. 1989. Proc. Int. Symp. on rice production on acidic soils in the tropics held at LF.S. Kandy, Srilanka.

Gaur, A.C., Neblakantan, S. and Dargan, K.S. 1984. Organic Manures. ICAR, New Delhi. DO 159.

Gaur. A.C. and Mukheriee. D. 1980. PI. Soil 56: 273.

George, Mercy and Prasad, R. 1989. Fert News 34(5): 21.

Ghosh, A. B. and Hasan, R. 1976. Bull Indian Soc. Soil Sci. 10:1.

Gilmour, C.M., Broadbent, F.E.,, Beck, S.M. 1977. Recycling of carbon and nitrogen through land disposal of various wastes. In: Eliott, L. F., Stevenson, F.J. (eds). Soils for management of organic wastes and wastewaters. ASA, CSSA. SSSA. Madison. DO 171-194.

Gnanamani, A. and R. Kasturi Bai. 1992. Influence of biodigested slurry on rice-gram cultivation. Bioresource Technology, 41:217-221.

Goel, P.K. and Trivedi, R.K. 1984. Some considerations on sewage disposal to fresh waters and resultant effects. Pollu. Res. 3: 7-12.

Goswami, B.K., Haque, M.M., Mishra, S.D. and Majumdar, V. 1988. Comparative efficiency of Hind-o-Meal, Karanj, Castor oil-seed cakes and carbofuran against root-knot nematode in cowpea. Ann. agric. Res. 10: 316» 318.

Goswarni, B.K. and Vijaylakshmi, K. 1986b. Efficacy of some

indigenous plant materials and non-edible oil seed cakes against Meloidogyne incognita on tomato. Indian J. NematoL 16: 280-281.

Govindaiah, Suryanarayanan, N., Sharma, D.D. and Gargi. 1989. Effect of mulching of green leaves for the control of root-knot nematode of mulberry. Indian J. Nematol. 19: 25-28.

Gupta, K. and Sinha, J.P. 1974. Indian J. Agric. Chem. 7: 25.

Handa, B.K. 1987. Nitrate content ofgroundwater in India. Pert. News 32: 11-29.

Harndi Y.A. 1982. FAO Soils Bull. No. 49, Rome. pp 188.

Harneed, S.F. 1970. Note on the effect of some organic additives on the incidence of root-knot nematodes in tomato. Indian J. Agric, Sci. 40: 207-210.

Hart, R.D. and Sands, M.W. 1991. Sustainable land use systems research and development. Presented at the USDA/ICAR/RRC Workshop on Sustainable Land Use Systems Research and Development, New Delhi, India,. Feb. 1990.

Hayman, D.S. 1982. Advances in Agricultural Microbiology, (ed). N.S. Sabba Rao, Oxford-IBH, New Delhi, pp 325-373.

Hide, G.A., Corbett, D.C.M. and Evans, K. 1984. Effects of soil treatments and cultivars on early dying disease of potatoes caused by Globodera rostochiansis and Verticillium dahlie. Ann. appl. Biol. 104: 277-289.

Hillel, D. 1987. The efficient use of water in irrigation. World Bank Technical Paper No. 64. The World Bank, Washington DC.

Idelovitch, E. 1978. Wastewater reuse bybiological-chemical treatment and groundwater recharge. J. Water Pollut Control Fed. 50:2723-2740

Janke, R.R. and Blew, J. 1992. Sustainable Land Use Systems Research in North America: An Assessment. Proc. International Workshop on Sustainable Land Use Systems Res., New Delhi, India, Feb. 12-16, 1990.

Jansson, S.L., Persson, J. 1982. Mineralization and immobilization of soil nitrogen. In: Stevenson, F.J. (ed). Nitrogen in agricultural soils. Am. Soc. Agron., Madison. Agronomy 22:229-252.

Jaradat, A.A. 1989. A progress report on cropping systems research in Jordan. ACSAD, pp 64.

Jaradat, A.A. 1992. Regional Assessment ofSustamable Land Use Systems Research. The Middle East and North Africa. Proc.

International Workshop on Sustainable Land Use Systems Res., New Delhi, India, Feb. 12-16, 1990.

Jewell, W.J., Madras, J. J., Clarkson, W.W., DeLancey-Pompe, H. and Kabrick, R.M. 1983. Wastewater treatment with plants in nutrient films. Report PB 83-247-494, US Environmental Protection Agency, Ada, Oklahoma.

Jeyabal, A., K. Thammathan, G. Kuppuswamy, and A.R., Lakshmanan. 1990. Effects of biogas slurry and other organic manures on root nodulation, growth and yield of groundnut cv. VRI 2 in vertisoL In Abstr. 30th Ann. Conf. Assoc. Microbiologists of India. Dep. Microbiol., Annamalai Univ., Annamalainagar, Tamil Nadu. P. 172.

Jha, K.P., Vamadevan, V.K., Dubey, A.N., Asthana, D.C. and Nanda, B.S. 1978.Oryza, 15: l.

John, P.S., Prasad, R„ Pandey, R.K. and Buresh, R.J. 1989. Pert. News 34(9): 19. c

Jones, C.A., Cole, C.V., Sharpley, A.N., Williams, J.R. 1984. A simplified soil and plant phosphorus model, 1: Documentation. Soil Sci. Soc. Am. J. 48: 800-805.

Jones, L.H.P., Jarvis, S.C. 1981. The fate of heavy metals. In: Greenland, D.J., Hayes, M.H.B. (eds). The chemistry of soil processes. John Wiley & Sons, New York, Chichester, pp 593-620.

Jurwarkar, A.S., Deshbharatar, P.B. and Bal, A.S. 1991. Exploitation of nutritional potential of sewage and sludge through land application. In: Asia Experiences in Integrated Plant Nutrition. RAPA- FAO Bangkok, pp 178-201.

Kaira, R.L. and Chawla, R.P. 1981. Monitoring of pesticides residues in the Indian Environment In: B.V. David, (ed). Indian Pesticide Industry-Facts and Figures. Vishvas Publishers, Bombay, pp 251-285.

Kandiah, A. 1990b. Criteria and classification of saline water. Water, Soil and Crop Management Practices Relating to the Use of Saline Water. A. Kandiah (ed.). AGL/MIC/16/90. FAO, Rome. 34-51.

Kansal, B.D. 1992. Effect of waste disposal on soils and plants. In: G.S. Dhaliwal, B.S., Hansraand N. Jerath, (eds). Changing Scenario of Our Environment. PAU, Ludhiana, pp 197-206. *

Kass, D.L., Burgos, C. and Vargas, A. 1992. An Assessment of Research on Sustainable Land Use Systems in Central America. Proc. International Workshop on Sustainable Land Use Systems Res.,

New Delhi, India, Feb. 12- 16, 1990.

Kathpal, T.S. and Kumari, B. 1993. Pesticide contamination of non-fatty foods in developing countries. In: G.S. Dhaliwal and B. Singh (eds). Pesticides: Their Ecological Impact in Developing Countries. Commonwealth Publishers, New Delhi, pp. 162-189.

Katyal, J.C., Holt, L.S. and Gadalla, A.M. 1989. A method to determine soil entrapped denitrification products in submerged soils. Soil. Sci. Soc. Am. J. 53:296-297.

Keeney, D.R. 1989. Sources of nitrate to groundwater. In: R.F. Follett, (ed). Nitrogen Management and Groundwater Protection. Elsevier, Amsterdam, pp 23-34.

Kemmier, G. 1986. Fert. News 31(3): 29.

Keren, R., Bingham, F.T. 1985. Boron in water, soils, and plants. Advances in soil science. Springer, Berlin Heidelberg, New York, pp 229-276.

Keren, R., Bingham, F.T., Rhoades, J.D. 1985. Effect of clay content in soil on boron uptake and yield of wheat. Soil Sci. Soc. Am. J. 49:1466-1470.

Ketkar, C.M. and Ketkar, M.S. 1993. Versatile neem. A source for plant protection. In: M.S. Chari and G. Ramaprasad, (eds). Botanical Pesticides in Integrated Pest Management. Proc. natn. Symp., Jan. 21-22, 1990, CTRI, Rajahmundry, pp 118-124.

Kostial, K. 1986. Cadmium. In: W. Mertz, (ed). Trace Elements in Human and Animal Nutrition. Academic Press, London, pp. 319-325.

Kumar, R. 1992. The extent ofNi build up in polluted soils and evaluation of its toxicity to plants. M.Sc. Thesis, PAU, Ludhiana.

Kumaraperumal, K., Baskaran, T.L., Charles Daniel, S., Rajamonickarn, D., Rajan, S.D. and Srinivasan, T.R. 1975. Indian Sugar, 25:681.

Kuppuswamy, G., A. R. Lakshmanan and A. Jeyabal. 1991. Effect of gypsum-enriched biogas sludge and farmyard manure on rice yield. Int. Rich Res. News. 16 (4): 18.

Kuppuswamy, G„ A. Jeyabal and A.R. Lakshmanan. 1992. Effect of enriched biodigested slurry and FYM on growth and yield of rice. Agric. Sci. Digest. 12(2): 101-104.

Kute. S.B. and Mann, H.S. 1968. Indian J. Agron. 13:20.

Ladha, J.K., Watanabe, 1. and Saono, S. 1988. Nitrogen fixation by leguminous green manure in rice farming. IARI, Los Banos,

Phillipines. pp 165.

Landau, M. and Tucker, J.W., Jr. 1984. Acute toxicity of EDB in aldicarb to two young esturian fish species. Bull. Environ. Contam. Toxicol. 33: 127- 132.

Lau, S.L., Ekern, P.C., Loh, P.C.S., Young, R.H.F. Burbank, N.C., Dugan, G.L. 1975. Water recycling of sewage effluents by irrigation: a field study in Ohau. Water Res. Cent., Univ. Hawaii, Honolulu, Tech. Rep. 79:110 pp.

Loria, R., Epiea, R.E., Baier, J.H., Martin, T.M. and Moyer, D.D. 1986. Efficacy of sweep-shank fumigation with 1,3- dichloropropene against Pratylenchus penetrans and subsequent ground water contamination. Pl. Dis. 70:42-45.

Mahapatra, I.C., Krishnan, K.S., Singh, M. and Jain, S.C. 1974. Pert. News 19(12): 17.

Maheshwari, R.C. 1984. Agricultural residues as a source of energy. In: Proc. Bioenergy Soc. 1st Conv. Symp. New Delhi, pp. 127-131.

Majid, F.Z. 1983. Aquatic weeds and algae, the neglected natural resources of Bangladesh (Booklet), 26 p.

Majid, F.Z., Lutfun Nahar, Sajeda Begum, Rahima Khatun and Nilofar Akhtar. 1980. Use of naturally occurring algae as biofertilizer or organic manure. National Workshop on algal systems proceedings, Madras, India, Paper No. 26, 11-14.

Mann, M.S., Takkar, P.N., Bansal, R.L. and Randhawa, N.S. 1978. J. Indian Soc. Soil. Sci. 26: 208.

Mann, M.S., Takkar, P.N., Bansal, R.L. and Randhawa, N.S. 1978. J. Indian Soc. Soil Sci. 26: 208.

Manning, E.W. 1986. Planning Canada's Resource Base for Sustainable Production in Canadian Agriculture in a Global Contest, Opportunities and Obligations, (ed). Knell, I.S. and English, J.R. Univ. of Waterloo Press, Waterloo, Ontario, Canada, N2L 3GI. pp 53-68.

Mariakulandi, A. and Morachan, Y.B. 1964. Madras agric. J. 51: 127.

Maynard, D.N., Barker, A.V., Minoti, P.L., Peck, N.H. 1976. Nitrate accumulation in vegetables. Adv. Agron. 28:71-118.

Mendis, M.S. 1988. A technical and economic cross evaluation of biomass residue densification systems. International Workshop on BM fuel Briquetting in Developing Countries, Khartourn, Sudan, Oct. 23-26, 1988.

Middlebrooks, E.J., Middlebrooks, C.H., Reynolds, J.H., Watters, G.Z., Reed, S.C. and George, D.B. 1982. Wastewater Stabilization Lagoon Design, Performance and Upgrading. Macmillan, New York.

Milman, A. 1990. Alanto- and isoalanto-lactones. Chem. Nat. Comp. 26: 251-262.

Miyazaki, Akira and Kada, Ryohei. 1992. Development and Present Issues of Sustainable Land Use Systems in Japan. Proc. International Workshop on Sustainable Land Use Systems Res., New Delhi, India, Feb. 12-16, 1990.

Mongia, A.D., Gangwar, B., Singh, S., Kumar, V. and Bandyopadhyay, A.K. 1989. Pert. News 34(4): 81.

Mortvedt, J.J. and Osborn, G. 1982. Studies on the chemical form of cadmium contaminants in phosphatic fertilizers. Soil Sci. 134: 185-192.

Moses Onim, J.F. 1992. Assessment of Sustainable Land Use Systems Research in East Africa. Proc. International Workshop on Sustainable Land Use Systems Res., New Delhi, India, Feb. 12-16, 1990.

Mukheljee, D. and Gaur, A.C. 1984. Ag. Res. Comm (Life Sci. Adv. Suppi.) 3:9.

Mukheljee, D. and Gaur, A.C. 1985. Indian Agric. 29:93.

Muralikrishna Reddy, V. 1987. Effect of biogas slurry and treated ureas on rice cv. ADT 36. M.Sc. (Ag.) Diss. Annamalai Univ. annamalainager, Tamil Nadu.

Nagar, S. 1975. Biogas plant based on night soil and animal dung. IRCWD News 1446.

Nagpal, N.K. 1985. Long-term phosphorus sorption in a brunisol in response to dosed effluent loading. J. Environ. Qual. 14: 280-285.

NPC (National Productivity Council). 1987. All India Report on Improvement of Agricultural Residues and Agro-industrial By-products Utilisation. N. Delhi.

NRC (National Research Council- ed). 1980. Mineral tolerance of domestic animals. Nati Acad. Sci., Washington, 577 pp.

NRC (National Research Council). 1992. Neern: A Tree for Solving Global Problems. Report of Adhoc Panel of the Board on Science and Technology for International Development, National Academy Press, Washington, DC.

Olsen, S.R., Sommers, L.E. 1982. Phosphorus. In: Page, A.L., Miller,

R.H., Keeney, D.R. (eds). Methods of soil analysis, 2: Chemical and microbiological properties. Am. Soc. Agron., Madison. Agronomy 9: 403- 430.

Page, A.L. 1974. Fate and effects of trace elements in sewage sludge when applied to agricultural lands. A literature review study. US Environ. Protect Ag. EPA-670-2-74-005.

Page, A.L., Chang, A.C., Sposito, G. and Mattigod, S. 1981. Trace elements in wastewater: their effects on plant growth and composition and their behaviour in soils. In Iskander IK (ed). Modelling wastewater renovation, land treatment. John Wiley & Sons, New York, pp 182-222.

Palazzo, A.J. 1977. Land application of wastewater: forage growth and utilization of applied nitrogen, phosphorus and potassium. In: Loehr, R.C. (ed). Land as a waste management alternative. Science Publisher, Ann. Arbor, pp 171-180

Panda, N. and Sahoo, D. 1989. Pert. News 34(4): 39.

Pandey, A. and Sushil Kumar. 1989. J. Sci. & Indust. Res. 48: 134-144.

Panse, V.G., Abraham, T.P. and Leelavathi, C.R. 1965. Green Manure Bull No. 2, ICAR, New Delhi.

Paoletti, M.G. 1992. Sustainable Agroecosystems in Western Europe. Proc. International Workshop on Sustainable Land Use Systems Res., New Delhi, India, Feb. 12-16, 1990.

Pareek, R.P. and Gaur, A.C. 1973. Curr. Sci. 42:278.

Pathak, B.S., Jain, A.K. and Bhatnagar, A.P. 1990. Fuel characteristics of some biomass species of Northern India. In: Proc. Int. Symp. on Application of Energy in Agriculture- Role of Biomass Fuels. Schools of Energy Studies for Agriculture, PAU, Ludhiana.

Pescod, M.B. 1986. Integration of urban wastewater treatment and effluent re-use in irrigation. Water Pollut. Control Fed. 85:459-471

Pettygrove, G.S., Asano, T. (eds). 1985. Irrigation with reclaimed municipal wastewater- aguidance manual. Lewis, Chelsea.

Prasad, M. and Prasad, R. 1983. Pl. Soil 70: 287.

Prasad, R., Devakumar, C. and Shivay, Y.S. 1993. Significance in increasing fertilizer nitrogen efficiency. In: N.S. Randhawa and B.S. Parmar, (eds). Neem Research and Development. Society of Pesticide Science, India, New Delhi, pp 97-108.

Purvis, D. 1985. Trace element contamination in the environment, rev. edn. Elsevier, Amsterdam.

Rana, R.L. and Kansal, B.D. 1983. Sorption and release of cadmium in sewage irrigated soils. J. Indian Soc. Soil Sci. 31: 439-443.

Rao, M.M. and Sharma, K.C. 1978. J. Indian Soc. Soil Sci. 26: 44.

Reddy, K.R. and DeBusk, W.F. 1987. Nutrient storage capabilities of aquatic and wetland plants. Aquatic Plants for Water Treatment and Resource Recovery. K.R. Reddy and W.H. Smith (eds). Magnolia Publishers, Orlando, Florida.

Ryden, J.C., Pratt, P.F. 1980. Phosphorus removal from wastewater applied to land. Hilgardia 48(1): 1-36.

Sadanandan, N. and Mahapatra, I.C. 1973. J. Indian Soc. Soil Sci. 21: 173.

Sahu, S.K. Nandu, 1988. Direct and residual effect of biogas residue application with nitrogen on rice yield. Int. Rice. NewsL 13(4): 28.

Sankaram, A. (1996). Eco-fanning Relevance and Rewards. Survey of Indian Agriculture, 1996. The Hindu, p. 35-37.

Sawhney, B.L. 1977. Predicting phosphate movement through soil columns. J. Environ. Qual 6: 86-89.

Saxena, M.C. and Siddiqui.M.K.J. 1982. Pesticide pollution in India: Organochlorine pesticides in women, buffalo and goat milk. J. Dairy Sci., 65: 430-434.

Saxena, R.C. 1993. Scope of neem for developing countries. In: Souvenir World Neem Conf, Feb. 24-28, 1993, Bangalore, pp 30-36.

Sehgal, V.K. and Bhatnagar, A.P. 1980. Agricultural wastes and their utilisation. Paper presented in XVII Annual Convention of ISAE, IARI, New Delhi, Feb. 6-8, 1980.

Sen, S. and W.C. Bonde, 1962 Effect of time of application ofFYM at different levels J. Indian soil Sci. 10:61-63.

SESA (School of Energy Studies for Agriculture). 1992. Agricultural wastes estimates. PAU, Ludhiana

Sharma, K.N. and Meelu, O.P. 1975. J. Indian Soc. Soil Sci. 23:76.

Sharma, R.C., Grewal, J.S. and Singh, M. 1979. J. Indian Soc. Soil Sci. 27:161.

Sharma, R.K. and Tiagi, B. 1989. Studies on efficacy of certain leaf powders against root-knot nematode on pea. Indian J. Mycol. PI. Pathol. 19: 101-102.

Sharma, S.N. and Prasad, R. 1974. Indian J. Agric. Sci. 44:229.

Shelef, G. 1977. Water reuse in Israel. In: Shuval, H.I. (ed). Water renovation and reuse. Academic Press, New York, London, pp

311-332.

Shuval, H.I. (ed). 1977. Health considerations in water renovation and reuse. In: Water renovation and reuse. Academic Press, New York, London, pp 33- 72.

Shuval, H.I. 1987. Recycling wastewater for agricultural use: regulations and control of health aspects. Paper prepared for the Land and Water Development Division, FAO, Rome.

Shuval, H.I., Adin, A., Fattal, B., Rawitz, E. and Yekutiel, P. 1986. Wastewater irrigation in developing countries: health effects and technical solutions. Technical Paper No. 51. World Bank, Washington DC.

Siddiqui, M.A. and Alam, M.M. 1990. Sawdusts as soil amendments for control ofnematodes infesting some vegetables. Biol. Wastes 33: 123-129.

Singh, B. and Dhaliwal, G.S. 1993. Pesticide contamination of fatty foods in developing countries. In: G.S. Dhaliwal and B. Singh (eds). Pesticides: Their Ecological Impact in Developing Countries. Commonwealth Publishers, N. Delhi, pp. 131-161.

Singh, G.B., Singh, K.N. and Singh, R.P. 1992. Assessment of Sustainable Land Use Systems Research in South Asia. Proc. International Workshop on Sustainable Land Use Systems Res., New Delhi, India, Feb. 12-16, 1990.

Singh, J. and Kansal, B.D. 1985a. Amount of nutrients and heavy metals in the sewage water of different towns of Punjab and its evaluation for irrigation. J. Res. PAU. 22: 17-24.

Singh, J.B., 1989. Biogas slurry manure. Moving Technology. 4:34-36.

Singh, K. and Yadav, D.V. 1985. Twenty five years of soil research on sugar crops, IISR, Lucknow, pp 26.

Singh, K.P., Pandey, S.Y. Khan, M.M. and Singh, Satendra, 1983. Residues ofaldicarb in coconut. Pestic. Sci. 14: 441-443.

Singh, P.K. 1988. Biofertilizers: Potentialities and Problems, (eds). S.P. Sen and P. Palit. Plant Physiol. Forum, Calcutta, pp 109-144.

Singh, R.P. 1993. Botanical pesticides in developing countries: Current status and future trends. In: G.S. Dhaliwal and B. Singh, (eds). Pesticides: Their Ecological Impact in Developing Countries. Commonwealth Publishers, N. Delhi, pp 236-269.

Singh, S. and Kanwar, R.S. 1964. Ind. J. Sugarcane Res. Develop. 8:388.

Sinhababu, S.P. and Sukul, N.C. 1990. Essential oils as nematicidal principles. Envir. Ecol. 8:1118-1120.

Smith, S.J., Schepers, J.S. and Porter, L.K. 1990. Assessing and managing nitrogen losses to the environment. Adv. Soil Sci. 14: 1-43.

Srimannarayana, G. 1993. Chemical investigation of indigenous insecticidal plants including neem and evaluation of insecticidal/ antifeedant activity of the constituents. In: M.S. Chari and G. Ramaprasad, (eds). Botanical Pesticides in Integrated Pest Management. Proc. natn. Symp., Jan. 21-22, 1990, CTRI, Rajahmundry, pp 62-73.

Sriskandarajah, N. 1992. Assessment of Sustainable Land Use Systems Research in the Oceania Region. Proc. International Workshop on Sustainable Land Use Systems Res., New Delhi, India, Feb. 12-16, 1990.

Stevenson, F.J. 1982. Humus Chemistry: Genesis, Composition, Reactions, Wiley-interscience, New York.

Stevenson, F.J. 1985. Cycles of soil, carbon, nitrogen, phosphorus sulfur, micronutrients. John Wiley & Sons, New York.

Swaminathan, M.S. (1995). Rational of integrated Systems. Survey of Indian Agriculture, 1995. The Hindu p. 7-15.

Talukder, N.M., M.S. Ali and A. Latif. 1988. Effect of biogas effluent 'on the yield and quality of rice. Int. Rice Res. Newsl. 37:11-16.

Tandon, H.L.S. 1989. Fert. News 34(4): 21. 190. Taylor, A.W., Kilmer, V.J. 1980. Agricultural phosphorus in the environment. In: Khasawneh, F.E., Sample, E.G., Kamprath, E.J. (eds). The role of phosphorus in agriculture. Am. Soc. Agron., Madison, pp 545-557.

Taylor, A. W., Kilmer, V.J. 1980. Agricultural phosphorus in the environment. In: Khasawneh, F.E., Sample, E.C., Kamprath, E.J. (eds). The role of phosphorus in agriculture. Am. Soc. Agron., Madison, pp. 545-557.

Thomas, R., Law, J.P. 1977. Properties of wastewaters. In: Eliott, L.F., Stevenson, F.J. (eds). Soils for management of organic wastes and wastewaters. SSSA ASA CSSA, Madison, pp. 47-72.

Tilak, K.V.B.R. 1987. Sci. Rep. 3: 17-19.

Tiller, K.G. 1989. Heavy metals in soils and their environmental significance. Adv. Soil Sci. 9: 113-142.

Timmer, L.W. and Leyden.R.F. 1980. New Phytol. 85: 15-23.

Tinker, P.B. 1980. Soil and Agriculture, (ed). P.B. tinker. Blackwell:

Oxford. pp 1-34.

Tiwari, K.N., Pathak, A.N. and Hari Ram, 1980a. J. Indian Soc. Soil Sci. 28: 162.

Tiwari, K.N., Pathak, A.N., Upadhyay, R.L., Hari Ram, Shukia, B.R. and Tripathi, S.K. 1979. Bull. Indian Soc. Soil Sci. 12: 519.

UNDP, 1989. Increased productivity of barley, pasture and sheep in Iraq, Jordan and Syria. RAB/87/021.

Venkatramani, G. (1995) Ecological farming- A viable option for future Survey of Indian Agriculture, 1995. The Hindu p. 23-31.

Venkatramani, G. (1996). Eco-technology towards intensive farming. Survey of Indian Agriculture, 1996. The Hindu p. 29-33.

Vimal, O.P. and Tyagi, P.D. 1984. Bioenergy R & D for developing countries. In: Proc. Bioenergy 1st Conv. Symp. New Delhi, pp 271-281.

Vyas, M.K. and Motiramani, D.P. 1971. J. Indian Soc. Soil. Sci. 19:39.

Weiss, R.F. 1981. The temporal and spatial distribution of tropospheric nitrous oxide. J. Geophys. Res. 86: 7185-7195.

Wheler, W.B., Thompson, M.P., Edelstein, R.L. and Krause, R.T. 1979. Ultrasonic extraction of carbofuran residues from radishes. Bull. Envir. contam. Toxicol. 21:238-242.

WHO (World Health Organisation- ed). 1973. Reuse of effluents. Methods of wastewater treatment and health safeguards: report of a WHO Meeting of Experts. WHO Geneva Tech. Rep. Ser. 517.

WHO. 1989. Health guidelines for the use of wastewater in agriculture and aquacuiture. Technical Report No. 778. WHO, Geneva 74 p.

Williams, C.H. and David, D.J. 1973. The effect of superphosphate on the cadmium content of soils and plants. Aust. J. Soil Res. I 1: 43-56.

Wright, D.J., Birtle, A. J., Crops, A.E. and Dybas, R.A. 1983. Efficacy of avermectins against a plant parasitic nematode. Ann. Appl. Biol. 103: 465-470.

Zhengfang, Li and Vouxia, Li. 1992. Assessment of Sustainable Land Use System Research in China. Proc. International Workshop on Sustainable Land Use Systems Res., New Delhi, India, Feb. 12-16, 1990.

Zodhy, L.L., S.M.S. Badr El-Din, M.A. Khalah and H.K. Abd. E.L. MakNud. 1984. Evaluation of digested slurry as a fertilizer. Egypt. J. Microbiol. 19:131-138.